D0850699

WITHDRAWN

PARABLE BEACH:
A PRIMER IN COASTAL ZONE ECONOMICS

The MIT Sea Grant Ocean Engineering Series
Dean Horn, general editor

1. Norman J. Padelford, Public Policy for the Use of the Seas

2. J. Harvey Evans and John C. Adamchak, Ocean Engineering Structures

3. Martin A. Abkowitz, Stability and Motion Control of Ocean Vehicles

4. Philip Mandel, Water, Air, and Interface Vehicles

5. Koichi Masubuchi, Materials for Ocean Engineering

6. John Craven, Ocean Engineering Systems

7. Norman J. Padelford, New Dimensions of U.S. Marine Policy

8. Dennis W. Ducsik, editor, Shoreline for the Public: A Handbook of Social, Economic, and Legal Considerations Regarding Public Recreational Use of the Nation's Coastal Shoreline

9. John W. Devanney, G. Ashe, and B. Parkhurst, Parable Beach: A Primer in Coastal Zone Management

PARABLE BEACH:
A PRIMER IN COASTAL ZONE ECONOMICS

J. W. Devanney III, G. Ashe, and B. Parkhurst

The MIT Press
Cambridge, Massachusetts, and London, England

PUBLISHER'S NOTE

This format is intended to reduce the cost of publishing certain
works in book form and to shorten the gap between editorial preparation and final publication. Detailed editing and composition
have been avoided by photographing the text of this book directly
from the author's typescript.

Library of Congress Cataloging in Publication Data

Devanney, John W
  Parable Beach.

  Bibliography: p.
  1.  Coastal zone management--United States.
I.  Ashe, G., joint author.  II.  Parkhurst, B.,
joint author.  III.  Title.
HT392.D48    333.9'17'0973        76-29350
ISBN 0-262-04052-2

To ourselves

CONTENTS

1
INTRODUCTION                                                    1

2
THE CASE OF THE PARABLE BEACH HIGHRISE                          3
The Town of Parable Beach                                      3
Paragon Park                                                    7
The Next Meeting                                               10

3
SOME NECESSARY BACKGROUND                                      12
The Concept of Real Municipal Income                          12
The Black Box Concept                                          13
The Implications of Accepting Market Prices as a
Measure of Value                                              14
Present Value                                                  16
Choice of Interest Rate                                        18
Inflation                                                     19
The Crucial Importance of Net Rather Than Gross              19

4
ANALYSIS OF THE PRODEVELOPMENT AND ANTIDEVELOPMENT
ARGUMENTS                                                     27
The First Step--A Set of Accounts                            27
The Second Step--Choice of a Baseline                        28
Analysis of the Town Hall Account                             28
Δ Town Hall--Daffyland                                        36
The Other Accounts                                            38
The Parable Beach Customers' Account                          40
The Parable Beach Apartment Tenants' Account                 42
The Parable Beach Boatowners' Account                        42
Putting It All Together                                        44
The Effect of Respending                                      50

APPENDIX TO CHAPTER 4
THE TREATMENT OF PRICE/QUANTITY CHANGES                       52
The Parable Beach Boatowners' Account                        52

5
THE ORONOCO REFINERY PROPOSAL                                 55
One More Into the Breach                                       55
George Banks's Report                                          56
The Labor Accounts                                            62
The Parable Beach Consumers' Account                          64
The Parable Beach Landowners' Account                        67
Summary                                                       70

6
THE ORONOCO REFINERY FROM THE POINT OF VIEW OF THE STATE     75
The Governor's Problem                                        75
George's Analysis for the State                               75

The Refinery Owners' Account                                    77
The State Oil Consumers' Account                                80
Putting It All Together Again                                   84

7
THE REFINERY FROM THE POINT OF VIEW OF THE NATION              88
The U.S. Construction and Operating Employees                  89
The Town Hall and State House Accounts                         90
The U.S. Property Owners' Account                              90
The U.S. Consumers' Account                                    90
The Refinery Owners' Account                                   91
The Federal Government Account                                 92
The Respending Account                                         94
Putting the Whole Thing Together a Third Time                  94

8
A CHECKLIST                                                    98

ACKNOWLEDGMENTS

This report was prepared for the U.S. Department of Commerce,
National Oceanic and Atmospheric Administration, Office of
Coastal Zone Management, under Contract No. 3-35484.

The report's publication in this form has been sponsored by
the M.I.T. Sea Grant Program, under Grant No. 04-5-158-1
from the Office of Sea Grant, National Oceanic and Atmospheric
Administration, U.S. Department of Commerce, as part of the
Program's goal of disseminating widely pertinent information
on ocean and coastal affairs.  The U.S. Government is authorized
to produce and distribute reprints for governmental purposes
notwithstanding the copyright notation that appears here.


An M.I.T. Sea Grant Program Related Report
Report No. MITSG 75-11
Index No. 75-811-Nde

PARABLE BEACH:
A PRIMER IN COASTAL ZONE ECONOMICS

Coastal zone management is a resource allocation problem. American society must somehow decide how to allocate an essentially fixed supply of coastal zone resources among growing public and private demands for coastal areas. Historically, the answer has been to allow supply and demand to determine the usage of coastal areas through the price mechanism--the use which paid the highest price for a particular property obtained it. Zoning provisions, public ownership, and tax laws have all had an impact on the market results, but the current allocation is essentially the result of private market operations.

Increasingly, these results have been called into question. A series of laws have been passed which attempt to modify the private market operation. Usually, these laws involve a transfer of at least part of the allocative decision to some public body. At present, along most of our coastline it is impossible to effect even a moderate-sized development without the approval of a number of municipal and state agencies. A large development will require the approval of a score of municipal, state, and federal bodies. This transfer of the allocative decisionmaking to public bodies places a heavy responsibility on the individuals within these bodies, for it is they who must now decide on how society uses its coastal zone. On any such decision, they will be besieged with arguments pro and con. The intensity of these pressures reflects the increasing value that society places on the coast and the subsequent importance of their decisions.

Often the arguments for and against some change in coastal zone allocation will take an economic form. This is a reflection of the fact that an extremely important measure of a town's or a state's well-being is its wealth--its ability to consume market goods. A prospective developer will claim that his proposal will have a substantial effect on the economy of the region and will buttress this claim with a great deal of analysis and figures. Similarly, anti-development forces will offer counterclaims also supported by extensive figures, expert testimony, and analysis.

The purpose of this manual is to aid the responsible public decisionmaker faced with such claims to sort out these arguments. We will be pointing out the common fallacies in many "economic" arguments, both pro and con a development, sifting out the truth, aiming to put the decisionmaker in a position to assess the true impact of a development on the market wealth of the political entity to which he is responsible, be it a town, a state, or the entire country. An important byproduct of our prescriptions for these public decisionmakers is the establishment of the true economic

basis of the conflict between individual municipality, state,
and country which is central to much of the coastal zone
issue.

Our procedure will be by allegory.  We will consider the
case of a hypothetical coastal community, Parable Beach, and
two important decisions which it finds itself faced with.  We
hope you will find Parable Beach an interesting place.

## Chapter 2

## THE CASE OF THE PARABLE BEACH HIGHRISE

### The Town of Parable Beach

The two women and five men who serve on the town council of
Parable Beach face a number of difficult decisions.  These
decisions are not made any easier by the fact that each member of the council is a volunteer with limited time to devote
to the council's activities.  None of them has any formal
training in regional economics, yet they are about to be
besieged with a variety of "economic" arguments and counter-
arguments on issues which they all realize will have a very
substantial impact on their town's future.

Parable Beach is a community whose year-round population
is a little over 10,000.  The town is situated on a narrow,
low-lying peninsula spitting out into Metropolitan Bay (see
Figure 2.1).  Parable Beach is located approximately twenty-
five land miles away from Schrod City, a large metropolitan
complex.  Its major asset is its three miles of oceanfront
shoreline, which features a large beach attracting a summer
population of about 30,000, and some 75,000 day visitors
from the Schrod City area on a hot summer weekend.

The town contains 2,000 acres, the bulk of which is situated at the root of the peninsula.  The peninsula itself
averages a little over a quarter-mile in width.  In the central portion of the root is an abandoned quarry covering
some 650 acres, presently being used as the town dump.

Despite the magnificent beach and the long bayfront,
Parable Beach's history has not been entirely happy.  At the
turn of the century, Parable Beach was a favored summer home
area for the wealthy and near-wealthy of Schrod City.  However, over the first third of this century, Parable Beach
fell out of favor with the rich.  The automobile made newer
coastal communities, further away from the city, accessible
to the wealthy, who found the large crowds of day visitors
traveling by excursion boats from the city on the weekends
incompatible with their desires for quiet and prestige.
Parable Beach fell out of fashion.

Residential construction all but stopped.  Presently
seventy-four percent of all housing in Parable Beach was
built prior to 1939, with eighty-five of that standing before
World War I.  Portions of the beachfront sprouted penny
arcades and beer joints serving the day people.  This trend
was aggravated by the fact that in 1912 the State Parks Com-
mission, in a legislative deal that is still being quarreled
about in the town, was able to gain control of the southern
third of the beach.  Parable Beach not only lost a rather
large piece of real estate from the tax rolls but, to add
insult to injury, found itself assessed a portion of the
costs of maintaining and protecting the State Parks Commis-
sion area.  On a ten-acre  site behind the public beach, an

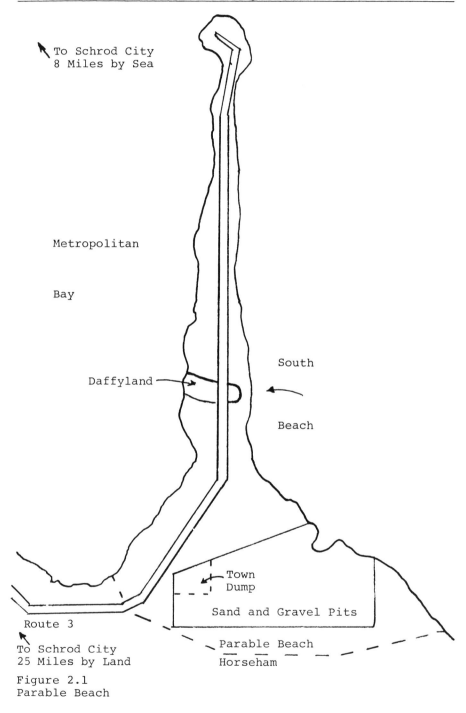

To Schrod City
8 Miles by Sea

Metropolitan

Bay

Daffyland

South

Beach

Town
Dump

Sand and Gravel Pits

Route 3

To Schrod City
25 Miles by Land

Parable Beach
Horseham

Figure 2.1
Parable Beach

amusement park, Daffyland-by-the-Sea, was erected, complete with roller coaster and carnival booths.

During this fifty-year period, the major transportation developments in the region tended to bypass the Parable Beach peninsula, making commuter access to Schrod City difficult. Only one rail line served the area, and this was abandoned before World War II. Since the area lacks any type of expressway, rush-hour travel time to the central business district of Schrod City by private car is an hour or more. A commute by public land transportation, involving two transfers, requires at least two hours. Therefore, middle-class commuters have not found Parable Beach a suitable home.

Abandoned by the rich and ignored by the middle class, after World War II Parable Beach became increasingly a workingman's town. Larger summer home properties were subdivided into very small single-family plots, many less than 5,000 square feet. Smaller properties were converted to year-round residences. Year-round population grew rapidly, as shown in Figure 2.2, while summer population fell slowly.

The new year-rounders were young, lower-middle-class whites. In 1970 only twenty-five percent of the families had incomes over $15,000, as opposed to thirty percent for the metropolitan region. In that year, average family income ($11,979) and median family income ($10,677) were lower than for the metropolitan area. Forty-four percent of the population is under twenty, as compared to thirty-five percent for the region. Even more striking is the fact that thirty-nine percent of the population is under eighteen. Some forty percent of the town's year-round population is in school. The rise in the enrollment of the Parable Beach public schools is depicted in Figure 2.3. This explosion in school-age population occurred at a time when unit education costs were also spiraling. Presently, Parable Beach's education budget is $3.8 million, or over $1,100 per student after federal aid.

Despite the relatively low incomes, most Parable Beach residents (eighty-nine percent) live in single-family homes, and most own their own homes. However, the median number of people per dwelling unit is 3.4 in Parable Beach, as opposed to 3 for the metropolitan area. Only four percent of these single-family homes were valued at $20,000 or more in 1970, as opposed to twenty-nine percent for the metropolitan area. The median value of such units in Parable Beach was $18,600 as compared to $21,800 for the entire area. Only eighteen percent of these homes have been built since 1950, and only 1.5 percent since 1965.

In short, the transition from a wealthy resort area to a lower-middle-income bedroom community plus metropolitan beach has not been easy for Parable Beach. It has, in fact, entailed a number of problems: rising cost of government far outstripping increase in property values, overcrowding and congestion, inadequacy of sewerage and schools. Now, however, the region as a whole has begun to take another look at Parable Beach, its magnificent beach and its relative proximity to town, at least as the crow flies. Even the abandoned

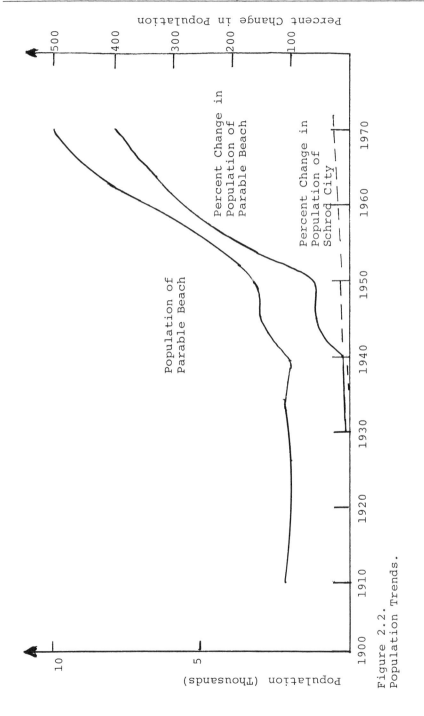

Figure 2.2.
Population Trends.

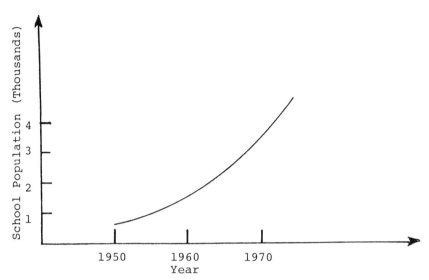

Figure 2.3.
School Population Trends.


quarry has drawn developer interest.  These development proposals present Parable Beach with a number of important
choices and the town is attempting to decide what to do with
them.

Paragon Park
The first proposal we want to consider is that of the Ithaca
Development Corporation.  Ithaca believes that the population
pressures on the metropolitan area are so great and the at-
traction of a smog-free, coastline location so large that it
can now sell high-rise apartments to middle-income commuters
despite the horrendous commute to Schrod City.  Ithaca has
identified a twelve-acre plot in the middle of the beach area
and obtained an option to buy the parcel.  Ithaca also envi-
sages the possibility of marina facilities, a moderate-sized
motel for summer patrons, and an associated commercial development to service the complex and replace existing business
which would be displaced.  To this end, it has approached
the town council for a variance to present town ordinances
with respect to building height and density.  The council
has called a meeting at which Ithaca has been asked to present its proposal in some detail.
    The council meeting is about to begin.  The Parable Beach

High School auditorium is full.  A few latecomers are still
trickling in the rear door, forcing the other standees outward
along the rear wall.  Simon D. Montfort, the chairman of the
council, calls the meeting to order.  He's a dentist who works
in Schrod City.  His family has lived in Parable Beach for
three generations.  After review and acceptance of the minutes
of the last meeting, Simon announces that the ordinary business of the council will be put off so that they can move
immediately to the Ithaca proposal.  After cautioning the
audience on the need for order, he introduces the head of the
Ithaca presentation team, Edwin Steuben, president of Ithaca.
Most of the people in the audience know Steuben from his pic-
tures in the Parable Patriot, the local weekly.  The Patriot
has run two rather unflattering articles on Steuben and Ithaca
and has taken a strong editorial stand against the development.  Steuben is a tall, grey, impressive man, but he reads
his presentation in an uncomfortable monotone, only occasion-
ally glancing up from his text.
    "Thank you, Dr. Montfort.  We at Ithaca are extremely
grateful to the council for granting us this opportunity to
present our ideas to Parable Beach.  As you know, I am not a
resident of this area.  However, almost every summer I have
sailed along this coast and I have never failed to marvel at
your beach and your location.  It is not surprising to find
that Parable Beach was one of the earliest permanent settle-
ments along this coast, a thriving trading station between
Indian and colonist.  It is easy to imagine that these
earliest settlers naturally assumed that Parable Beach, with
its unique location, magnificent beach, and beautiful hills,
would be the hub around which future development would take
place, the focal point, the apex of regional development.
    "We all know that that dream did not come true.  It is
perhaps in the nature of smaller communities that they go
through cycles, responding to the ebb and flow of society's
sometimes capricious whims.  For the last fifty years, the
region has not treated Parable Beach kindly.  The state has
taken one-third of your beach and in return assessed you for
its upkeep.  Parable Beach has been ignored by the state's
transportation planners and as a result it is more isolated
from Schrod City, eight miles away, than towns fifty miles
inland.
    "However, we at Ithaca believe the tide is turning.  The
same unique features which attracted the Indian and the early
settler are still here.  However, it is up to us to capitalize
on them.  As a first great step in this direction, Ithaca
proposes Paragon Park."
    The hall lights darken and an artist's rendering of the
development is projected on a screen behind the council.
    "Paragon Park is an integrated residential, retail, and
recreational development carefully planned to take full advantage of a unique site.  Paragon Park would occupy a twelve-
acre area currently containing Daffyland.  Paragon Park will
be built in three phases.  The first phase will involve the
construction of 262 modern oceanfront apartments in two

fifteen-story towers, together with forty-two units of du-
plex garden apartments. The second phase involves a 500-slip
marina and shopping center on the bayfront, and the third
phase will involve an additional three towers containing some
400 apartments.

"What would Paragon Park mean for Parable Beach? In
purely monetary terms, Paragon Park will involve a total expenditure of close to $20 million. At present tax rates,
this development when completed will generate $1.5 million
per year in property taxes. This is equivalent to a ten percent increase in the tax base or an average decrease in annual property taxes of $125 for each and every property owner
in Parable Beach.

"And this is only the beginning. The construction phase
will require the expenditure of $12 million on local labor.
Many of these jobs will go to Parable Beach residents, inject-
ing this money into the town's economy. Further, construction personnel will make considerable expenditures in the
area. Our consultants estimate that six percent of the payroll, or $720,000, will be spent in the area of the site,
generating revenues for local businesses and retailers. To
this must be added the expenditures of the new residents of
Paragon Park. Independent economic consultants tell us that
some ten percent of their income, estimated at $9 million,
will be spent in the immediate area. Thus, some $900,000
will be added to the town's revenues. Further, these expenditures will generate respending, which will have an additional impact on the local economy. This is the 'multiplier
effect,' as what one person spends is spent again and again.
Our experts inform us that the effect of this respending
will be to increase the direct impact of this spending by a
factor of 3.5. In summary, then:

| | |
|---|---:|
| Annual property tax revenue | $1,600,000 |
| Spending by construction | 720,000 |
| Spending by residents | 900,000 |
| Multiplier effect | 5,000,000 |
| Total annual impact on local economy | $8,000,000." |

Steuben then proceeds to a more detailed description of
the physical development, finally ending with the prediction
that "Ithaca fully expects Paragon Park to become the corner-
stone on which a new era for Parable Beach will be built."

Montfort thanks Mr. Steuben, and notes that he will be
available for questioning later, but indicates that in the
interest of making the issues involved as clear as possible,
he first wants to call on Mr. Giles D. Ray, head of the South
Beach Businessmen's Association.

"Thanks, Simon. It's my honor to be representing the
South Beach Businessmen's Association tonight but, in a larger
sense, I and the SBBA are representing all the people of
Parable Beach whose destiny you and the rest of the council
hold in your hands.

"We have heard a most impressive presentation about what

is essentially a high-density urban housing project which the
developers wish to transplant into our coastal suburban community.  Tonight I'm going to leave to others the assessment
of what this transplantation of Schrod City will mean for
Parable Beach in terms of scenic values, community texture
and our whole way of life.  Tonight I'm going to concentrate
on the economic impact of this development on the town.  In
a word, economically this development would be a disaster
for Parable Beach.

"Presently, the largest export industry in Parable Beach
is Daffyland-by-the-sea.  Each year over 500,000 visitors
come to Parable Beach.  SBBA records indicate that on the
average each of these visitors spends $4.45 for a total reve-
nue of $2.25 million.  The Ithaca proposal would involve the
demolition of Daffyland together with its thirty-three concessions, but also the destruction of eighteen businesses on
properties surrounding Daffyland.  These businesses have a
total gross annual income of $650,000.  Together with Daffy-
land, they employ 167 people and have a gross payroll of
$800,000.  Destroy these businesses and you will have de-
stroyed over $2.9 million of direct revenues, $800,000 in
payroll and $400,000 in property taxes, for a total of $5.1
million.  And these are only the direct effects.  What about
the indirect effects as the businessmen of South Beach re-
spend this money, much of it in Parable Beach, and that money
is respent and so on?  I would say that Mr. Steuben's multiplier of 3.5 is conservative.  In any event, even if the multiplier were only 1.6, we would be talking about losses in
excess of $8 million, the best we can expect from the new
development.  At a multiplier of 3.5 we are talking about a
loss of $17 million per year.

"In summary, the proposal involves the destruction of
fifty-one local businesses including the largest export industry in the town, the loss of 167 local jobs, and a gross
loss to the local economy which is likely to exceed $10 mil-
lion.  In accepting Paragon Park, you will be cutting off
the economic blood system upon which Parable Beach is based."

A lively discussion follows, which after a time tends to
move away from the strictly economic impact and concentrate
on scenic values, congestion, and, in a disjointed manner,
on the community's view of itself.

At 11:30 Dr. Montfort adjourns the meeting after the
council agrees to meet in a closed session on the following
Monday.

The Next Meeting
It is a week after the Ithaca presentation and the town council is meeting in camera.  Dr. Montfort has attempted to fo-
cus the discussion for this meeting on the economic arguments
offered by the prodevelopment and antidevelopment forces.
The ensuing discussion is confused and increasingly heated,
consisting mainly of comments on the accuracy of individual
numbers.  Two of the seven council members are clearly for
the development; two others, against.  And the comments on

the numbers gradually become comments on the veracity of the
sources of the numbers.

Finally, Kathleen O'Houlihan, the oldest member of the
council, takes the floor.

"I for one wouldn't give two ants for either side's figures.  They just seem to pile things on top of things.  Take
this multiplier effect:  if what is paid to me makes me rich
and when I turn around and pay it to another, that makes him
rich, why does it ever stop?  Why isn't the multiplier ten
or one hundred or one thousand?  And where are the costs to
the town of the high-rise--the police protection, fire protection, sewerage, highway maintenance and schooling?  Where,
for that matter, are the costs to the town of Daffyland, the
police, the parking, the congestion?

"I don't believe the figures in their present form are
even worth discussing.  I believe we need professional help.
George Banks, who lives three doors down from me, works in
town at the State Office of Coastal Zone Management.  I've
talked to him and he has agreed to look at these figures and
see what they really mean for the town's economic well-being.
I strongly move we take him up on his offer, turn over all
the data we have to him, and see what we come up with."

Mrs. O'Houlihan's suggestions meet with little real approval, but after another couple of hours of fruitless wrang-
ling, the council decides they have little to lose and
authorizes her to turn over the data to Banks.  It is agreed
that Banks will give his report to the council in two weeks.

## Chapter 3

## SOME NECESSARY BACKGROUND

### The Concept of Real Municipal Income

Let us digress from our story for a moment to think carefully about the problem which the town council faces. The council realizes that the market wealth of the town of Parable Beach-- the amount of market goods which the town as a whole can consume--will almost certainly not be the same with Daffyland occupying the South Beach site as it will with Paragon Park. At this point, however, the council hasn't the foggiest idea whether the town's wealth will be higher or lower with the high-rise rather than with the amusement park. Yet it must make a decision. The council realizes that municipal wealth is only one measure of a community's well-being. It also realizes that it is an extremely important measure. The council knows the people of Parable Beach are extremely concerned about the admittedly poor quality of their schools despite rapidly escalating tax rates. The people of Parable Beach are not wealthy. Almost all the homes of Parable Beach would welcome a little more income and most could ill afford a little less. Therefore, it is imperative that the council have a realistic estimate of the difference in municipal income associated with opting for Paragon Park rather than Daffyland. Once it has such an estimate, it will be in a position to balance this change in market wealth against the differences in environmental quality and scenic values which the alternatives imply before reaching a final decision.

In order to sort out the dollar figures offered by the prodevelopment and antidevelopment forces, we must first define just what we mean by the *real municipal income of Parable Beach.*

Perhaps the easiest way of getting at our definition of real municipal income is to imagine that the town of Parable Beach is owned and controlled by a single personage--Uncle Eph we might call him. Uncle Eph is interested in the total value, at present market prices, of all the goods he can consume with the output of the rather extensive resources he controls. Uncle Eph realizes that he can allocate his resources in an infinite variety of ways, some of which will allow him to consume a higher total value of goods than others. Uncle Eph, for reasons he chooses not to discuss, would like to make this market value of his consumption as large as possible.

His resources include not only the land and water, the buildings and roads, vehicles and vessels of Parable Beach, but also its present human inhabitants. We might regard this latter brand of resource as Uncle Eph's fingers, in that they both produce and consume. Uncle Eph has no particular feelings about his fingers. He isn't interested in whether one finger rather than another consumes a greater share of

the total value of all the goods he consumes.  He is only
interested in the total.  He considers himself better off if
this total value is larger, worse off if it's smaller, re-
gardless of the distribution of production and consumption
among his fingers.
   *We define the total value of the goods, priced at current
market prices, which Uncle Eph can consume, to be the real
municipal income of Parable Beach.*[1]
   Notice that in attempting to maximize this quantity, Uncle
Eph is ignoring the fact that any proposed change in the allocation of his resources will almost certainly make some of
his fingers worse off and some better off.  Uncle Eph simply
doesn't care.  He prefers the change if the total value of
the consumption of all his fingers is higher after the change
than before.  He will eschew the change if the total value is
less.  *Our concept of municipal income ignores the distribu-
tional effects of any proposed change within the town.*
   This limitation has obvious political implications, for
what may be a net increase to the town as a whole can affect
a particular set of losers quite adversely.  For example,
real municipal income will be increased by a change which
increases the real income of ninety percent of the town's
citizens by ten percent and decreases the real income of one
percent of the town's population by seventy percent, virtu-
ally wiping out this latter group.  Almost all urban renewal
schemes have had some such effects.  The carnies of Daffyland
could be this type of victim for the proposal under consideration.  Uncle Eph doesn't care, as long as he comes out ahead
overall.
   There is another thing to notice about Uncle Eph.  His is
a provincial and basically selfish character.  He only cares
about his own ability to consume.  He is completely indifferent to any effect, up or down, his choices might have on the
income of entities outside the town--the rest of the state,
for example, or the rest of the country.  Any change in income, no matter how large, to someone who is not a citizen
of Parable Beach is given no weight at all by our concept of
*municipal* income.
   Thus, in order to implement this parochial philosophy, we
will have to be quite precise about what we mean by a citizen
of Parable Beach.  Let us assume that the council is willing
to accept all *present* property owners in Parable Beach, both
year-round and summer, and all present year-round tenants as
citizens of Parable Beach.  Notice that this definition
applies to the situation *before* development.  Changes in income to people who are drawn into Parable Beach by the development will not be counted under this definition.

## The Black Box Concept
One way of thinking about this is to imagine that we enclose
the town in a black box.  We count the changes in real income,
up and down, to everybody presently within that black box
------------------------------------------------------------
[1]Throughout the book, "current" is taken to be 1974.

and ignore the changes outside the black box.

It should be obvious that in a very real sense the decision to place our black box around the town of Parable Beach
is arbitrary.  We could have drawn the black box around any
political entity.  Whether or not the town council of Parable
Beach cares to admit the fact, there are human beings outside
of Parable Beach whose real incomes will be affected by *some*
of the council's decisions.  And even if the council is not
concerned with these impacts, other, broader portions of society are.

In order to analyze the conflicts which necessarily arise
from these varying levels of responsibility, in this handbook we will actually be concerned with three different black
boxes (see Figure 3.1):

1.  A black box around the town of Parable Beach;
2.  A black box drawn around the state in which Parable
    Beach is located;
3.  A black box drawn around the entire nation.[2]

In Chapters 5, 6, and 7 we will be analyzing the same
development from the point of view of real municipal income,
from the point of view of real state income, and from the
point of view of real federal income.  This will mean keep-
ing three separate sets of accounts, and in general we will
obtain three different answers.

For now, the important point to draw from this is:  whenever one is analyzing the economic effects of a development,
*one must make one's black box explicit*--whose income are you
talking about?  Failure to do so is endemic to the coastal
zone development debates and results in the meaningless tos-
sing around of numbers--a process which has already started
in Parable Beach.

In our calculations, we will also be dealing with developer real income, not so much because of our basic interest
in the developer's well-being, but because an increase in
his income will be necessary to induce the developer to under-
take the project under analysis.  For these calculations, we
will simply draw our black box around the developer.

## The Implications of Accepting Market Prices as a Measure of Value

We have already discussed two of the limitations of real income as a measure of well-being--it does not account for
changes in income distribution within the black box and it
ignores changes outside the black box.  There are some more
fundamental limitations.  In using real income as a measure
of well-being, one is accepting the market's valuation of
all goods.

There is some confusion about the philosophical implica-
------------------------------------------------------------

[2]Other choices of black box are possible:  for example, the
SMSA in which Parable Beach is located.  However, since most
of the political power is focused at these three levels,
these are the three most interesting black boxes from the
point of view of the coastal zone.

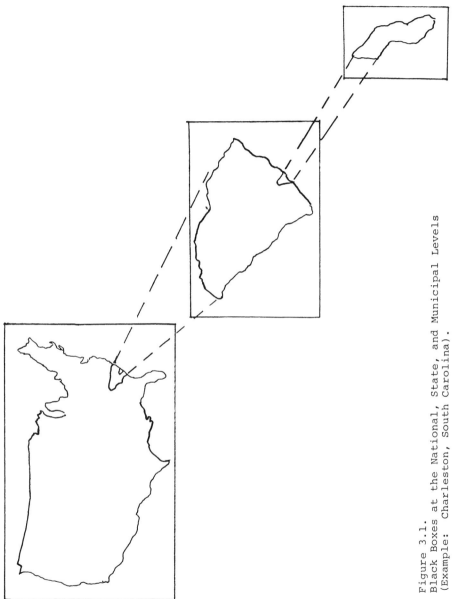

Figure 3.1.
Black Boxes at the National, State, and Municipal Levels
(Example:  Charleston, South Carolina).

tions of accepting this valuation--to many it implies approval of the market's valuation of, say, a carton of cigarettes
as equal in value to a symphony ticket.  But it's less a matter of approval than the acceptance of the following reality:
for most smaller black boxes--state, municipality, indivi-
dual--*most market prices are for all practical purposes fixed*.
These entities cannot in general affect prices.  Given that
market prices are fixed, there is a clear indication that,
*everything else being equal*, people would rather be able to
consume more than less.  This is not necessarily a crass attitude.  Ability to consume in this context covers consumption of education, ballet, and museum memberships, as well
as consumption of golf carts, television sets, and beer.
Rather, accepting real income as a measure of well-being involves taking a laissez-faire attitude towards how the entity
within the black box chooses to spend its income.  Thus, if
one feels that people's consumption patterns are self-
destructive, one will be quite unhappy with real income as a
measure of well-being.

It is also true that, insofar as the prevailing prices
are set in competitive markets, these prices reflect the relative willingness to pay of the society's inhabitants for the
goods so priced.  Thus, *if one accepts the present income
distribution*, one can make an argument that these prices are
an imperfect but indicative reflection of people's underlying
desires.  Further, even substantial changes in income distribution are likely to have little effect on most prices.

This argument holds only for those goods for which a func-
tioning market exists.  The fact that no market exists in
which water quality or air quality or scenic values can be
exchanged means that these goods can be severely underpriced
in terms of society's underlying willingness to pay.  In this
situation, the fact that the market sets the price of air
quality at zero to someone who would use (diminish) this air
quality does not necessarily imply that the people in the
society are not willing to pay anything to avoid this loss
in air quality.  Such environmental issues are ignored by
real income.

In short, real income is only one dimension of an entity's
well-being, albeit an extremely important one.  The goal of
this manual is to see that the analysis regarding this dimension, at least, is fallacy-free and truly informative.  We
will not deal directly with other measures of well-being,
such as environmental quality, simply because it is necessary
to take one thing at a time.  We must learn to walk before
we can run.

## Present Value
Before we can even walk, we must face one more additional
problem associated with changes in income--whether municipal,
state, or federal income--and that is:  changes in income
can occur at different points in time.  The effects on municipal income, both positive and negative, of changing Daffyland to a high-rise housing complex will be felt at varying

points in time ranging from almost immediately to perhaps
fifty years in the future.  Paragon Park will involve construction expenditures over the next few years but little or
no property tax payments in this period.  Later on, after
the investment is complete, property tax and expenditures by
residents will grow.  Daffyland promises a more even set of
financial flows.  How are we to balance income effects now
with income effects further, and in some cases much further,
into the future?  To see, let us take our black box for the
moment to be Parable Beach and look at the problem from the
point of view of Uncle Eph.

Uncle Eph is a shrewd old codger.  He realizes that there
is considerable difference between receiving one dollar in
additional income now and one dollar in additional income
ten full years from now.  The reason, of course, is that
Uncle Eph has the opportunity to invest the one dollar received now at some annual interest rate, say ten percent.
After one year so invested, Uncle Eph will have $1.10, which
he can reinvest for a second year, obtaining an additional
ten percent on $1.10, or eleven cents, for a total of $1.21,
which he can reinvest, and so on.  If he invests the dollar
received now for ten years at ten percent, he will find that
at the end of the tenth year, his investment will be worth
$2.59, which is quite different from one dollar.  The timing
with which he receives the same amount of additional municipal income obviously makes a great deal of difference to
Uncle Eph.

To put it another way, if Uncle Eph has investment opportunities which can earn him ten percent per year, receiving
one dollar now is equivalent to receiving $2.59 ten years
from now.  He would be indifferent between receiving one dollar now and $2.59 ten years from now but he would certainly
not be indifferent between receiving one dollar now and one
dollar ten years from now.

Uncle Eph, therefore, realizes he has to put increases in
municipal income received at varying points in time on a common temporal basis.  He chooses to relate them to an equivalent amount received now (1974).  That is, in valuing an
increase of one dollar which will occur ten years from now,
he asks himself what is the amount received now which will
grow to one dollar ten years from now.  Mathematically he
is asking:

What number x $2.59 = $1.00?

The number he is after is simply $1.00/$2.59 or 38.6¢.  This
number is called the *present value* of a sum of one dollar
received ten years from now assuming a ten percent interest
rate.  In general, the present value of a sum $x_n$ received n
years from now at an interest rate i is

$$\frac{x_n}{(1 + i)^n}$$

If we are dealing with a development alternative which will
increase municipal income by $x_1$ in year 1, $x_2$ year 2, and so
on through N years, then the present value of all these increases, V, is simply the sum of the present values of each
yearly increase or

$$V = \frac{x_1}{(1 + i)} + \frac{x_2}{(1 + i)^2} + \frac{x_3}{(1 + i)^3} + \dots \frac{x_N}{(1 + i)^N}$$

Uncle Eph reasons that, given his opportunity to reinvest at
an interest rate i, he would be just as well off in terms of
his real wealth if he received the sum V now as if he received
the entire stream of future increases in income resulting
from the development alternative.  Thus, in comparing various
development alternatives, he will do so on the basis of their
present values, that is, on the basis of an equivalent amount
of income received in 1974 on a one-shot basis.

The justification for applying Uncle Eph's reasoning to
our valuations of Parable Beach's alternatives follows from
the fact that the individual citizens of Parable Beach are
either borrowers or lenders or both.  In so far as they are
lenders, they are in exactly the same position as Uncle Eph
and therefore future income must be adjusted downward relative to present income according to the interest rate at which
he can lend.  In so far as a citizen who could be a lender
does not do so, he is making a clear statement that he prefers $1.00 worth of consumption now to $1.00(1 + i) worth of
consumption a year from now.  In both cases, future increases
in income must be present valued at the interest rate available to these citizens.  By the same token, to the extent
that the citizens of Parable Beach are borrowers, they are
making a clear statement that they are willing to exchange
($1.00 + i) a year from now for a dollar's worth of consump-
tion now where i is the interest rate at which they are borrowing.  The town itself as a municipal entity is a borrower.
Obviously, an additional dollar of net public revenues received now could be used to pay off a portion of the town
debt which will cost ($1.00 + i) to pay off a year from now.
In all these cases, a dollar received now is worth more than
a dollar received a year from now by an amount which depends
on the relevant interest rate.  Putting things in terms of
present value, that is, equating future increases in income
to an equivalent amount received now, accounts for these differences.  Therefore, in evaluating the alternatives before
Parable Beach, we must estimate the net increases or decreases
in municipal income in each year associated with the alternative and then convert this time stream of revenues and outlays to an equivalent amount received now by means of present
value.

## Choice of Interest Rate
Our use of present value raises the problem of how to choose
the interest rate we use in our analyses.  Clearly, some of

the inhabitants of Parable Beach have different investment
and borrowing opportunities from others.  From the point of
view of the town as a whole, what we require is a weighted
average of the highest rates being paid by individual borrowers and the lowest rates being obtained by individual lenders where the weights would depend on the degree to which
each individual is affected by the development alternatives.
Unfortunately, accurately estimating such an average is a
hopeless task.  Fortunately, it's also usually unnecessary.
As we shall see, it is usually possible to obtain all the
insight we require by running our analyses over a range of
interest rates and examining the results in aggregate.  When
an estimate is required, a ballpark value can often be obtained by using the highest rate of interest being paid by
the town on its borrowings.

## Inflation
At this point, we had better say a word about inflation.
All our analyses are based on 1974 prices.  Thus, for example,
if a particular policeman's services were priced at $5.00 an
hour in 1974, we will assume that his wage is $5.00 an hour
in 1984.  In reality, the general price level may have risen
so that *in 1984 figures* the policeman is earning, say, $6.00
an hour in 1984.  However, we will implicitly deflate these
prices back to 1974 dollars to put everything on the same
basis.  This holds for all future prices and costs.  In particular, this procedure requires that we use inflation-free
interest rates in obtaining present values.  For instance,
if an investor's best employment of capital is to buy a bond
at a market interest rate of ten percent for a given period
during which price levels rise at three percent per year,
the investor will realize a seven percent growth in his income in real purchasing power (in constant value dollars).
Thus, in this book, when we speak of an interest rate of
eight percent, we are talking about eight percent net of inflation, which at present would correspond to a market interest rate of twelve percent or more.

## The Crucial Importance of Net Rather Than Gross
The final and in some ways the most important point to make
about a decision to measure municipal well-being in terms of
municipal income is that in assessing any two alternative
developments, the only thing that counts is the *net difference* in real income between the two within the black box
we've drawn around the municipality.  This is an obvious
statement but one that is frequently ignored in the public
debate concerning potential developments.
    To see what we are driving at, consider what is, in the
present context, a reasonably neutral example.  Our black
box is some well-defined region.  This region is contempla-
ting the construction of an office building on a particular
plot of land which is currently devoted to an intensively
cultivated truck farm.  The potential developers have determined to their satisfaction that the building is at least as

good an investment for their capital as they could make elsewhere. Having done this, they approach the region to convince it that it is in the region's interest to have this
project undertaken.

Potential developers usually concentrate on the *input*
side, that is, with regard to the resources which will be
employed in their proposal. In this case, they note that
the building will cost them $10 million, of which half will
be spent within the region. They point out that the bulk of
this $5 million will be respent within the region; the re-
spent money will in turn be again respent; and so on. In
this manner, the initial expenditure will be "multiplied".
Further, upkeep and maintenance of the building will involve
expenditures of $500,000 a year; these expenses are also sub-
ject to respending. Depending on the multiplier assumed,
they conclude that the building will have an effect on the
region's economy several--and sometimes many--times larger
than the gross expenditures associated with the project.

Opponents of new developments generally argue on the *output* side, that is, with regard to the value of the goods the
current use of the land produces.[3] Thus, the opponents of
the proposed building point out that the gross revenues of
the truck farm are $2 million a year, most of which goes to
the farm's 200 workers who will lose their jobs. They argue
that not only will the region's economy lose $2 million a
year directly, but also farm equipment, fertilizer, and seed
suppliers and others throughout the region whose livelihood
depends on the respending of the farm workers will be adversely affected, starting a multiplier chain. The multipliers assumed by opponents of a development are usually of
roughly the same size as those of the proponents; the difference is that these multipliers are usually applied to the
gross value of the outputs of the present use of the site
rather than the gross value of the inputs.[4]

Both sides are making a basic error in talking gross
rather than net changes. Assuming that the developers' and
opponents' underlying figures are current, let's examine this
decision about resource allocation from the point of view of
our concept of black box income. That is, let's attempt to
determine the *difference* in real regional income (as defined
earlier) between constructing the office building and leav-
ing the site as a truck farm. Start with the developers' $5
million to be spent within the region. For simplicity, assume

-------------------------------------------------------------

[3] In reality, both opponents and proponents of new developments often mix input-side and output-side arguments. However, we can make our point using the above simplification
without loss in generality since, as we shall see, the same
basic principles refer to both lines of reasoning.

[4] It really doesn't make much difference, for assuming normal
profit levels, the gross expenses of any undertaking will be
approximately equal to the gross revenues.

this all goes to construction labor.  The first question we
have to ask ourselves is:  what would this labor be earning
if it weren't employed on this building?  The market value
of this alternative production measures *the cost to the
black box*--as opposed to the cost to the developer--of using
this resource on the building.[5]  If there is nothing else
this labor could be doing, obviously the alternative opportunity value of the labor is nothing.  The transfer of the
labor to this project involves no loss in regional production elsewhere and hence the cost to the region of employing
the labor on the building is zero.  In this case, the
full $5 million payroll represents a net increase in re-
gional income, as is evidenced by the increase in the pay
the construction workers actually take home.  But if the construction workers could be employed elsewhere producing something else whose value approximates the wage rate--the
situation when we have full employment--then the cost to the
region of diverting this labor to the proposed building is
the loss in output elsewhere.[6]  Under full employment
this loss would be about $5 million and the payroll in
itself generates *no* increase in regional income, as is evidenced by the fact that whether the construction worker works
on this particular building or somewhere else, his paycheck
is unchanged.  In the case of full employment the cost to the
developer of the labor, $5 million, and the cost to the re-
gion are approximately equal.  In this situation, the net
increase in regional income due to the construction payroll
is zero.

  *The net effect on black box income of expenditures on
black box inputs depends critically on the alternative opportunities for employment that these resources have.*  This cru-
cially important principle is rarely emphasized in the public
debate concerning proposed developments.  It applies to all
black box inputs:  land, capital, and materials as well as
labor.  For any such input, the direct effect of its purchase
on regional income can range from the full expenditure, if
there is complete unemployment of that resource, to zero if
the resource is fully employed, and indeed can even be negative if the market price of the input is lower than the market value of its output.[7]

---------------------------------------------------------------

[5]To see this, look at it from Uncle Eph's point of view.

[6]Full employment means no excess supply of labor, which in
turn implies that regional employers find that the wage rate
is low enough that the market value of what each laborer produces is at least as great as the wage rate.  Therefore,
under full employment, the market price of labor is no higher
than the value of what that labor can produce elsewhere.

[7]This happens when a good is priced at less than the opportunity value of its use to the region due to regulatory control (as with natural gas and foreign currency) or the lack

However, in applying this principle we will concentrate
on the input labor.  The reason for this is that it is ordinarily difficult for the market price of land or equipment
or fuel to rise much above their opportunity value for a variety of reasons:  mobility, antitrust legislation, ability to
be stored.  But the input labor is often in partial unemployment; that is, the opportunity value of labor is often somewhat less than the market price, though rarely is it a great
deal less.  Setting the wage rate too high above opportunity
cost generates unemployment, which generates people willing
to work for less than the prevailing wage; this willingness
puts a limit on how far above the opportunity cost of labor
the wage rate can rise.

Now let's turn to the multiplier.  If we have full employment and there is no change in construction workers' income
with and without the building, then the net effect on regional income of the multiplier effect is zero, since there
is no difference to multiply.  But let's say we are facing a
partial unemployment situation such that the direct effect
on construction workers' take-home is twenty percent of the
gross payroll, or $1 million.  Then we can properly apply a
multiplier *to this difference*, for there will be a difference in the workers' expenditures.

However, in so doing we must once again be careful to
obtain the *net effect* of this additional respending on regional income.  Let's suppose that as a result of his increase in income, the construction worker spends an additional
$5.00 a week on, say, clothing.  The clothing retailer sees
an increase in his gross revenues of $5.00.  Does that mean
there is a net increase in regional income of that amount?
Obviously not.  The worker's expenditures require the employment of certain resources:  the retailer's help, the labor,
and the capital required to manufacture the apparel and bring
it to market.  Even if all these resources are found inside
the region, the cost to the region is again the alternative
opportunity value of these resources--the value of what they
could be producing if employed elsewhere.  It is true that,
due to fluctuations in demand, service industries tend to
operate at less than capacity much of the time.  Thus, if
the worker happens to spend his extra clothes money at a
time when retail capacity is in oversupply, then the additional cost to the store owner and to the region of selling
these additional items is little more than the wholesale
cost of the clothes.  The rest is profit and this, less the
bulk of extraregional taxes, is a true increase in regional
income due to the increase in the store owner's net income.
If, on the other hand, the money is spent during the Christmas

--------------------------------------------------------------

of a market (as with air and water quality).  Such a situation is known as *rationing*, for at the market price the supply
of the good will be less than the demand.  Rationing (a shor-
tage) is the opposite of unemployment (a surplus).

rush, then either the store owner will be forced to hire additional help or the quality of service to the rest of his
customers will decrease (they will bear some of the cost).
If retail capacity was correctly set before the increase,
the store owner will find that his net income has increased
very little as a result of the additional expenditures.[8]  In
short, the same kind of partial unemployment we find in the
direct labor markets we also find in the markets in which
respending occurs.  The net increase due to first-round respending is some percentage of the actual expenditures; this
percentage depends on (a) the amount of black box labor input to the good or service; and (b) the degree of unemployment in the black box respending market.  For most respending
markets, twenty percent would be a generous estimate of the
fractional increase in black box income.  The same kind of
argument holds for the second round of respending (the store
owner's additional purchases of clothing and help) and so on
ad infinitum, except that if full employment obtains in any
of these markets or all the resources used are extraregional
then the chain is broken and the increases in black box income stop at that point.

    For the sake of argument, let us assume that the chain is
never broken, that goods in all markets are priced at twenty
percent more than the opportunity cost of their inputs to the
region.  What is the net effect of this entire multiplier
chain on black box income?  It is $1 million (first round,
twenty percent of $5 million) plus $200,000 (second round,
twenty percent of $1 million) plus $40,000 (third round) plus
$8,000 (fourth round) and so on.  If one adds up these rapidly
decreasing amounts for as many rounds as one wishes to consider, the total approaches $1.25 million or twenty-five per-
cent of the original $5 million payroll.  In short, the *net*
effect of the multiplier phenomenon on black box income is
generally much overstated.  Moreover, its influence drops
off rapidly in two or three rounds.  Multiplier effects are
rarely anywhere near as striking as commonly claimed.  Of
course, we knew this.  If the multipliers of three and five
that developers often claim actually affected black box income by that amount, then we would have invented the money
tree.  By simply undertaking more and more expenditures, we
could increase black box income indefinitely.  The developers'
arguments are based on the implicit assumption that the resources used by their project cannot be used in any other
way.[9]  This is rarely, if ever, the case.

---------------------------------------------------------------

[8]Or the store owner may respond to the increased demand by
raising prices, which will have no net effect on regional
income, for the increase in real income of the seller will
be matched by a decrease in real income of the buyer.

[9]More precisely, we must distinguish between the gross mul-
tiplier--the total amount of economic activity required to

The arguments against the proposed building must be examined from the same point of view: what is the difference in black box income? It is true, of course, that if the office building is constructed the region will lose the output of the truck farm, and the farm's total revenue is the *gross* market value of this loss. Does this mean that regional income will decrease by this amount? Not necessarily, for as the opponents themselves have pointed out, if the farm goes, several other things will happen as well. First of all, 200 farm workers will lose their jobs. This means that 200 more people will now be available for employment elsewhere. If we have full employment, the workers will, after a time, find employment elsewhere at wages (producing output whose value is) approximately what they were earning and the net effect on regional income due to the job loss will only be the difference in their pay during the transition period plus any *net* multiplier effect on this difference. Similarly, the land will find employment elsewhere as building space; the truck farm owners will not sell the land unless they feel they are at least as well off after the transaction as before.

In short, from the gross loss in output we must subtract the value of what the farm's workers will produce elsewhere, the resale value of the farm's equipment, and the payments to the farm's owners for the land to obtain the direct loss in regional income associated with the demise of the truck farm. In a perfectly competitive economy, this difference would be zero, both the landowner and the truck farmer being indifferent to the change. In actual fact, it's rarely zero; ordinarily the displaced labor will suffer a real loss in income, at least during a transition period, and the landowner will experience a gain. The net effect can be either positive or negative but it rarely approaches anything like the gross value of the former output.

To this difference we must apply a multiplier to account for partial unemployment elsewhere in the economy, but exactly the same line of reasoning holds for this multiplier effect as for the multiplier effects resulting from changes in construction workers' take-home pay. That is, the indirect change in regional income is a fraction applied to a fraction of the gross value of the output of the truck farm.

In short, the same kind of "let's look at the difference" viewpoint applies to the present use of the resources as well as the proposed new development, and the same kind of differential magnitudes obtain. Our concentration on the *net* effects on black box income of a proposed change is a two-edged sword biting deeply into *both* the usual "economic"

---------------------------------------------------------------

support a specified investment--and the net multiplier effect, which deducts from this total the value of the output of these resources in alternative employment. It is the latter concept that is relevant to regional income discussions.

arguments for development as well as the "economic" arguments
against.

When one takes this differential point of view in analyzing two alternative allocations of some resource, attention
necessarily becomes focused on those areas where the real
*changes* in black-box income generally reside, rather than on
the gross expenditure or gross revenues associated with the
various alternatives.  These areas are:

1.  The difference in the cost of the outputs to black box
consumers--will the alteration result in any changes in market prices within the black box?

2.  The difference in private profits to the black box investors affected;

3.  The difference in public profits (tax revenues minus additional cost of services occasioned by the developments under
consideration) to the black box public bodies affected;

4.  The difference in take-home pay to all the black box labor
affected;

5.  The net effect due to respending of all the above differences.

Notice that, when considering two alternative uses of a
resource, Items 4 and 5 generally are counterbalancing:  if
there is full employment, there will be little difference in
take-home pay of the labor involved under either alternative
and hence very little net multiplier effect.  On the other
hand, if there is extensive unemployment, then there may be
a sizable difference in take-home pay for the two sets of
laborers, but it will be a sizable difference for the laborers
involved in *both* alternatives.  If we build the office, the
truck farm labor suffers a sizable loss; if we keep the farm,
construction labor suffers a sizable loss.  The two individual losses may be substantial and the net effect on black
box income, their difference, still be small.  The same is
true of the multiplier effect.  With extensive unemployment,
it may be considerable for *both* the proposed development and
its alternative, but those two effects have to be subtracted
to obtain the net effect on black box income.  It's only when
  · there is substantial unemployment which is actually
reached by the development, and
  · one alternative employs little or no black-box labor resources and the other a lot,
that differences in black box income due to changes in labor
income and respending become noticeable on net.

Of course, this cancellation phenomenon is not going to
mollify either the farm workers or the construction unions,
who will continue to lobby vigorously for their respective
options.  They, unlike Uncle Eph, are quite concerned about
which fingers do the consuming.  The point is that this activity, however  vociferous, does not necessarily imply that
any net black box income is at stake.

By the same token, prodevelopment and antidevelopment economic arguments are generally not serious analyses, but rather sales pitches.  The Parable Beach town council is not

completely naive.  They know that neither Ithaca Development
Corporation nor the South Beach Businessmen's Association is
truly interested in the real municipal income of Parable
Beach.  These groups are interested in the real income of
Ithaca Development Corporation and the SBBA respectively.
Therefore, if the council wants an unbiased estimate of the
change in Parable Beach income associated with the high-rise,
it will have to make this estimate itself or obtain it from
a disinterested, competent third party.

## Chapter 4

ANALYSIS OF THE PRODEVELOPMENT AND ANTIDEVELOPMENT
ARGUMENTS

George Banks, at Mrs. O'Houlihan's urging, is attempting to
estimate the present-valued difference in Parable Beach real
income which would result from the town's opting for the Paragon Park development rather than the present use of the site.
He realizes his job is to count all the *changes* in income inside the Parable Beach black box *once and only once*. Banks
further realizes that this is easier said than done. From
sad experience, he knows that analyses which involve doubly
and triply counting certain effects while completely missing
others are the rule rather than the exception.

The First Step--A Set of Accounts
In order to avoid these pitfalls, it is extremely good prac-
tice--indeed necessary--to set up a comprehensive, consistent set of accounts to keep track of the various real income
changes within the black box currently under analysis. Usually, this can be done a number of ways. The most appropriate
choice of accounts will depend on the alternatives under analysis *and* the black box whose income is being analyzed.
    For his problem, Banks decides to break down the Parable
Beach income effects for each of the two alternatives as
shown in Table 4.1.
    The set of accounts must be collectively exhaustive and
mutually exclusive. That is, the accounts must include all
entities within the black box whose real income is likely to
be materially affected by the alternatives under consideration, and the accounts must be defined such that the same
effect is not counted twice. Thus, if a presently unemployed
Parablite will be hired as a policeman by the town for

Table 4.1
Individual Accounts for Paragon Park Analysis

| Paragon Park | Daffyland |
|---|---|
| Δ Town Hall* | Δ Town Hall |
| Δ Parable Beach construction labor | Δ PB employees |
| Δ PB employees | Δ PB concessionaires |
| Δ PB concessionaires | Δ PB customers |
| Δ PB customers | |
| Δ PB tenants--apartments | |
| Δ PB tenants--marine | |
| Δ Respending | Δ Respending |

*Throughout this paper, the Greek letter "delta" (Δ) is used
to signify change. Thus, "Δ Town Hall income" means "change
in town hall income."

Paragon Park with municipal revenues obtained from the development, one must not end up counting these municipal revenues twice--once when they are paid to the town and again when they are paid to the new policeman.  The straightforward way to do this is to keep careful track of *all* the revenues and *outlays* associated with the alternatives for each account. In other words, in the Town Hall account, count all the public revenues associated with the development, but from this deduct all the public outlays associated with the development, including the payment to the new policeman.  In the PB employees account, count the policeman's pay, and from this deduct what he would have been earning or receiving in state and federal welfare.  The various accounts should be chosen in such a way as to make this bookkeeping as easy as possible.

Some shortcuts are permissible and will not get the analysis into trouble as long as they are done carefully.  For example, the formerly unemployed policeman will *involuntarily* transfer a portion of his new-found income outside the black box in the form of federal and state income tax.  Thus, a true measure of his increase in income as a result of his new job is the difference in his pay after federal and state income tax and what he was receiving on welfare.  The strictly correct way of handling this is to show the income tax payments as an outflow in the employees account, but often we will be a little sloppy and work directly with the after-tax income in the employees account.

The Second Step--Choice of a Baseline
Our basic goal is to determine the *change* in municipal income resulting from opting for one development rather than another.  In order to do this, *it is extremely important that one work from a consistent baseline*--the situation against which changes are measured.  Sometimes it is most convenient to use the status quo or a carefully defined projection of the status quo as a baseline.  However, in situations where one is explicitly analyzing all the possible alternative uses of a particular site, it is often most convenient and instructive to work from a fictitious baseline--to assume that if the site is not used for the alternative currently under analysis, it will simply disappear.

The advantage of this fictitious construction is that with it one can analyze each possible alternative independently. Let us illustrate this procedure with the Paragon Park problem.

Analysis of the Town Hall Account
The first element in Table 4.1 accounts for the changes in the revenues and outlays of the Parable Beach body politic, i.e. those changes which are reflected in public revenues and expenditures for public services.  In the case at hand, George Banks has decided to assume the site must go to either Paragon Park or Daffyland.  There are no other possible uses. Under this assumption, as we shall see, it is permissible

and useful to generate the financial flows associated with
each alternative acting as if, were that alternative not undertaken, the site would disappear. Thus, in Table 4.2, which
shows Banks's estimates for the Town Hall account under the
Paragon Park alternative, Banks has put in the property tax
column his estimate of the *gross* property taxes which Paragon Park will pay in each year through the future. By the
same token, in the Δ Police column, the figures should be
the difference in Parable Beach police expenses between what
they would be with Paragon Park and what they would be if
the site disappeared, that is, the *additional* expenses occasioned by Paragon Park rather than nothing. Similarly for
fire, sewage, and street maintenance.

The idea is that Banks will operate under the same assumption in analyzing Daffyland. After we have completed similar analyses for all accounts and all alternatives, we can
obtain the *change* in real Parable Beach income by deducting
the total present value for all accounts under one alternative from the total under the other.

It is also possible to work with one of the alternatives
as a baseline. In that case, in each column of Table 4.2,
for example, the entry should be the difference in revenues
or outlays between Paragon Park and Daffyland. However, experience indicates that the fictitious baseline leads to
fewer errors, makes all the assumptions used in the analysis
more visible, and is a construction which lends itself better to delegation of the analysis of various alternatives to
a number of people. The important thing to keep in mind is
not the choice of a baseline, but rather, *whatever baseline
is chosen must be maintained consistently throughout the
analysis*. This is crucial.

Returning to Table 4.2, let's see how Banks arrives at
some of these estimates. His first step is to try to gather
from those involved in the development some figures which
would indicate the timetable of the development. After a
number of interviews with representatives of different fac-
ets of the construction industry, he is able to come up with
a rough estimate on which to base his calculations. This
timetable is shown in Table 4.3. Table 4.4 is a breakdown
of the individual phases.

The property valuation estimates shown in Table 4.3 are
based on conversations with the assessors of Parable Beach,
who indicate that the projects would probably be assessed at
fifty percent of present investment while under construction
and at ninety percent of replacement value after completion.
Then Banks, using the town's past tax reports, constructs
what he feels would be a realistic tax rate schedule for the
next ten years for three specific cases: with nothing on
the site; with the Paragon Park development; and with the
amusement park, Daffyland (see Table 4.5). Thus, using these
figures, it is a relatively straightforward matter to arrive
at yearly property tax bills for the Paragon Park developers.

The basic assumption Banks uses here is that Parable Beach
will wish to spend the same amount overall on public services

Table 4.2
Δ Town Hall Under Paragon Park Development

| | Revenues | | | | Outlays | | | | | Net |
|---|---|---|---|---|---|---|---|---|---|---|
| Year | Property Taxes | Sewerage Fees | Licensing Fees | Change in Cherry Sheet | Police/ Property Protection | Fire Protection | Sanitation Costs | Street Costs | School Costs | Net Cash Flow |
| 1974 | 5,990 | 0 | 0 | 0 | 0 | 0 | 0 | 0 | 0 | 5,990 |
| 1975 | 190,554 | 1,700 | 0 | 0 | 0 | 0 | -5,000 | -10,000 | 0 | 177,244 |
| 1976 | 406,499 | 2,200 | 0 | 0 | 0 | 0 | -5,000 | -10,000 | 0 | 393,699 |
| 1977 | 699,069 | 57,720 | 0 | 870 | -18,547 | -9,273 | -14,592 | -6,000 | -82,840 | 626,407 |
| 1978 | 1,053,608 | 88,420 | 7,200 | 896 | -28,824 | -19,146 | -14,592 | -1,500 | -82,840 | 1,002,812 |
| 1979 | 1,200,597 | 90,720 | 7,200 | 896 | -28,824 | -19,146 | -15,072 | -1,500 | -82,840 | 1,115,631 |
| 1980 | 1,590,310 | 92,620 | 7,200 | 896 | -28,824 | -19,146 | -15,072 | -3,000 | -82,840 | 1,510,664 |
| 1981 | 1,725,100 | 161,920 | 10,500 | 2,114 | -56,644 | -49,693 | -34,272 | -1,500 | -191,840 | 1,565,683 |
| 1982 | 1,810,501 | 163,420 | 10,500 | 2,114 | -56,644 | -49,693 | -34,272 | -10,000 | -191,840 | 1,644,084 |
| 1983 | 1,935,729 | 163,420 | 10,500 | 2,114 | -56,644 | -49,693 | -34,272 | -1,500 | -191,840 | 1,777,812 |
| 1984 | 2,016,033 | 163,420 | 12,000 | 2,144 | -56,644 | -49,693 | -34,272 | -10,000 | -191,840 | 1,861,615 |
| 1985 | 2,069,086 | 163,420 | 12,000 | 2,144 | -56,644 | -49,693 | -34,272 | -1,500 | -191,840 | 1,914,668 |
| 1986 | 2,157,509 | 163,420 | 12,000 | 2,144 | -56,644 | -49,693 | -34,272 | -1,500 | -191,840 | 2,003,091 |
| 1987 | 2,228,247 | 163,420 | 12,000 | 2,144 | -56,644 | -49,693 | -34,272 | -1,500 | -191,840 | 2,073,829 |
| 1988 | 2,281,300 | 163,420 | 12,000 | 2,144 | -56,644 | -49,693 | -34,272 | -1,500 | -191,840 | 2,126,882 |
| 1989 | 2,369,723 | 163,420 | 12,000 | 2,144 | -56,644 | -49,693 | -34,272 | -1,500 | -191,840 | 2,214,305 |
| 1990 | 2,422,776 | 163,420 | 12,000 | 2,144 | -56,644 | -49,693 | -34,272 | -1,500 | -191,840 | 2,268,358 |
| 1991 | 2,493,514 | 163,420 | 12,000 | 2,144 | -56,644 | -49,693 | -34,272 | -1,500 | -191,840 | 2,339,096 |
| 1992 | 2,546,568 | 163,420 | 12,000 | 2,144 | -56,644 | -49,693 | -34,272 | -1,500 | -191,840 | 2,392,150 |
| 1993 | 2,617,306 | 163,420 | 12,000 | 2,144 | -56,644 | -49,693 | -34,272 | -1,500 | -191,840 | 2,462,888 |
| 1994 | 2,688,044 | 163,420 | 12,000 | 2,144 | -56,644 | -49,693 | -34,272 | -1,500 | -191,840 | 2,533,626 |
| 1995 | 2,758,782 | 163,420 | 12,000 | 2,144 | -56,644 | -49,693 | -34,272 | -1,500 | -191,840 | 2,604,364 |
| 1996 | 2,811,836 | 163,420 | 12,000 | 2,144 | -56,644 | -49,693 | -34,272 | -1,500 | -191,840 | 2,657,418 |
| 1997 | 2,882,574 | 163,420 | 12,000 | 2,144 | -56,644 | -49,693 | -34,272 | -1,500 | -191,840 | 2,728,156 |
| 1998 | 2,953,312 | 163,420 | 12,000 | 2,144 | -56,644 | -49,693 | -34,272 | -1,500 | -191,840 | 2,798,894 |
| 1999 | 3,006,365 | 163,420 | 12,000 | 2,144 | -56,644 | -49,693 | -34,272 | -1,500 | -191,840 | 2,851,947 |

Table 4.3
Timetable for Development of Paragon Park

| Year | Action | Duration During This Year (Months) | Land Utilized (Acres) | Land Improved (Square Feet) | Costs for Land ($ x 10^{-6}$) | Construction Costs ($ x 10^{-6}$) | Individual Property Valuation ($ x 10^{-6}$) | Total Valuation ($ x 10^{-6}$) |
|---|---|---|---|---|---|---|---|---|
| 1974 | Acquire land for Project One* | 3 | 4.07 | 251,600 | 0.2664 | | 0.2398 | 0.0599 |
| 1975 | Begin construction, Project One | 12 | | | | | 1.5750 | 1.8148 |
| 1976 | Acquire land for Project Two<br>Begin construction, Project Two<br>Continue construction, Project One | 12<br>12 | 1.49 | 41,500 | 0.0945 | 0.73<br>3.14 | 0.0851<br>0.3650<br>3.1450 | 3.8359 |
| 1977 | Complete Project One<br>Continue Project Two | 12 | | | | 0.73 | 5.6610<br>0.6570 | 6.6578 |
| 1978 | Complete Project Two<br>Acquire land for Project Three<br>Begin construction, Project Three | 12 | 5.00 | 430,000 | 0.3259 | 5.2700 | 1.3140<br>0.2943<br>2.6350 | 10.2292 |
| 1979 | Continue Project Three | 12 | | | | 2.6300 | 3.9500 | 11.5442 |
| 1980 | Continue Project Three | 12 | | | | 2.6300 | 7.9050 | 15.4992 |
| 1981 | Complete Project Three | | | | | | 9.4860 | 17.0802 |
| 1982 | | | | | | | | 17.0802 |
| 1983 | Acquire land for Project Four<br>Begin construction, Project Four | 12 | 1.00 | 27,000 | 0.0585 | 0.6150 | 0.0508<br>0.3080 | 17.1310<br>17.4390 |
| 1984 | Complete Project Four | | | | | | 0.5535 | 17.6845 |

*See Table 4.4 for project composition.

Table 4.4
Composition of Individual Projects in Paragon Park
Development

| Project | Description | Land Utilized (Acres) | Total Cost ($ Million) |
|---|---|---|---|
| One | 262 unit high-rise apartment | 2.62 | 5.4119 |
| | 42 garden apartments | 1.45 | 1.1445 |
| Two | First increment, shopping center (15 stores) | 1 | 0.8445 |
| | Marina (500 slips) | 0.49 | 0.7100 |
| Three | Second increment, shopping center (10 stores) | 1 | 0.8445 |
| | 400 unit high-rise apartment | 4.00 | 10.0214 |
| Four | Third increment, shopping center (6 stores) | 0.34 | 0.2610 |
| | Motel (70 units) | 0.56 | 0.4125 |
| Project Total | | 11.46 | 19.6503 |

Table 4.5
Tax Rate Schedule Estimates for Varying Alternatives (Figures in Millions of Dollars; Tax Rate per $1,000)

| Year | Parable Beach Property Valuation Without Any Development | Tax To Be Raised | Tax Rate Without Any Development | Paragon Park | | Daffyland | |
|---|---|---|---|---|---|---|---|
| | | | | Total PB Valuation | Tax Rate | Total PB Valuation | Tax Rate |
| 1974 | 47.728 | 4.777 | 100 | 47.788 | 100 | 51.3227 | 93 |
| 1975 | 48.028 | 5.255 | 109 | 49.843 | 105 | 51.5927 | 102 |
| 1976 | 48.228 | 5.518 | 114 | 52.063 | 106 | 51.8227 | 106 |
| 1977 | 48.528 | 5.794 | 119 | 44.186 | 105 | 52.8227 | 106 |
| 1978 | 48.728 | 6.094 | 125 | 59.007 | 103 | 52.5217 | 116 |
| 1979 | 50.028 | 6.438 | 129 | 61.572 | 104 | 53.8517 | 120 |
| 1980 | 52.228 | 6.760 | 129 | 67.727 | 100 | 56.1837 | 120 |
| 1981 | 52.528 | 7.060 | 134 | 69.600 | 102 | 56.4537 | 125 |
| 1982 | 52.728 | 7.413 | 141 | 69.808 | 106 | 56.8337 | 130 |
| 1983 | 53.028 | 7.783 | 147 | 70.209 | 106 | 57.1787 | 136 |
| 1984 | 53.228 | 8.083 | 152 | 70.913 | 111 | 57.4237 | 141 |
| 1985 | 53.528 | 8.383 | 157 | 71.213 | 114 | 57.7267 | 145 |
| 1986 | 53.728 | 8.733 | 162 | 71.513 | 117 | 58.0287 | 150 |
| 1987 | 54.028 | 9.030 | 167 | 71.813 | 122 | 58.3289 | 155 |
| 1988 | 54.228 | 9.330 | 172 | 72.113 | 126 | 58.6292 | 159 |
| 1989 | 56.528 | 9.683 | 171 | 72.413 | 129 | 58.9292 | 164 |
| 1990 | 56.728 | 9.983 | 176 | 72.713 | 134 | 59.2122 | 168 |
| 1991 | 57.028 | 10.283 | 180 | 78.013 | 137 | 59.5152 | 173 |
| 1992 | 57.228 | 10.586 | 185 | 73.313 | 141 | 59.8182 | 177 |
| 1993 | 57.728 | 10.919 | 189 | 73.613 | 144 | 60.1185 | 182 |
| 1994 | 58.028 | 11.219 | 193 | 73.913 | 148 | 60.4188 | 186 |
| 1995 | 58.528 | 11.549 | 198 | 74.213 | 152 | 66.7488 | 190 |
| 1996 | 58.528 | 11.882 | 203 | 74.513 | 156 | 61.0491 | 195 |
| 1997 | 58.728 | 12.182 | 207 | 74.813 | 159 | 61.3791 | 198 |
| 1998 | 59.028 | 12.515 | 212 | 75.113 | 163 | 61.6791 | 203 |
| 1999 | 59.228 | 12.815 | 216 | 75.413 | 167 | 71.9824 | 207 |

with Daffyland as with Paragon Park.  Notice that it would
be double counting to count both the property tax paid by
Paragon Park and the resultant cut in tax rate to other Parable Beach property holders.  Notice also that Parable Beach
could increase its municipal income under the Paragon Park
alternative by increasing public services rather than cutting the tax rate.  Banks, however, assumes that the town
would take the increase in municipal income in the form of
a property tax cut.

The second column in Table 4.2 accounts for the income
the town would receive in installation, property betterment,
and monthly usage fees in connection with the town's central
sewage treatment system.  These figures are based on a six
dollar per foot installation charge, a six dollar per foot
betterment assessment, and a fifteen dollar per unit monthly
usage fee.  A unit is considered to be the equivalent of one
full-sized home bathroom and kitchen unit, or any fraction
thereof.

Banks uses previous years' licensing fees as relatively
reliable figures on which to judge the contribution in this
area that Paragon Park's commercial endeavors would make to
the town coffers.  With an estimate of twenty future stores
opening, as indicated in the timetable, and an average of
$330 per year in fees for a business person engaged in the
type of endeavor that would be planned for Paragon Park,
these figures, too, are straightforward to derive.

In the state in which Parable Beach is located, a portion
of the state revenues collected from each municipality is
returned to the town in a document known as the "cherry
sheet."  Banks makes a back-of-an-envelope calculation of
the state taxes paid by Paragon Park and takes a historical
fraction of this revenue to obtain the increase in the cherry
sheet.  Since the numbers come out small, more detailed analysis is not indicated.

On the outlay side, the areas involving police and fire
protection require some value judgments on Banks's part.
First, he assumes that with an average of 3.1 people per new
apartment and a one hundred percent increase in harbor use
resulting from the marina, the town will be forced to add
new employees as outlined in Table 4.6.  These estimates are

Table 4.6
New Municipal Employees Under Paragon Park Alternative

| Year | New Police | Summer-Only Police | New Fire-fighters | New Part-Time Firefighters | Average Salary |
|------|-----------|--------------------|-------------------|----------------------------|----------------|
| 1977 | 2 | | 1 | | $9,916/yr |
| | | | | 1 | $400/yr |
| 1978 | 1 (harbor patrol) | 2 | 1 | | $128/wk $9,916/yr |
| 1981 | 3 | | 2 | | $9,916/yr |
| | | | | 2 | $400/yr |

compared against crime rate and fire incidence increases in
the area as a function of population density and property
value, and Banks feels reasonably sure that they are accurate
enough for the purposes of his study.  Sanitation and street
cost figures are arrived at also from the analysis of the impact of undertakings of similar scale in the area, and Banks
finds that in the sewage treatment area, a five dollar per
linear foot installation cost and a four dollar per unit
treatment cost are representative.  By study of the archi-
tect's layouts for the total Paragon Park development and
the knowledge that street construction costs are running at
about ten dollars per foot while maintenance of the streets
planned for this area will run around $1,500 per year, Banks
is able to come up with an estimate for street costs associated with development.

Banks finds that the development would have a large impact on the school budget.  After talking with members of the
area's planning council, Banks decides that, in spite of what
the developers are saying, it would be realistic to expect
that ten percent of the new town inhabitants would be school-
age children.  This, combined with an average annual cost of
$1,090 per child for schooling, led to the entries shown in
Table 4.1.  A more accurate computation would involve estimating the cost of any *new* buildings and teachers required
by the *additional* schoolchildren.  If the town's school resources are underutilized, the unit cost of additional children will be lower, possibly much lower, than the average;
if the school system is operating at or above capacity, forcing new construction and hiring, the cost can be much larger
than the average.

Having generated all the town hall revenues and outlays
under the Paragon Park alternative in each year, Banks simply
sums them up to obtain the final column of Table 4.2.

The present value of the change in Town Hall income under
Paragon Park as opposed to the fictitious option "site disap-
pears" can be obtained by simply summing the elements of the
last column of Table 4.2 weighted by the proper compound interest factor.  Suppose we assume that the relevant interest
rate is ten percent real; then we have the sum

$$5,990 + \left[\frac{1}{1.10}\right]177,244 + \left[\frac{1}{1.10}\right]^2 393,699 + \left[\frac{1}{1.10}\right]^3 626,407$$

$$+ \text{ etcetera } \ldots + \left[\frac{1}{1.10}\right]^{25} 2,794,894$$

which equals $11.5 million.

Notice that Banks has cut the analysis off after twenty-
five years.  The present value factor for year 25 is

$$(1/1.10)^{25}$$

which is only .09, and those for ensuing years are still
lower.  Thus, it would take extremely large differences

between alternatives for income changes past year 25 to materially affect the overall present values. This fact allows the computation to be cut off at a moderate number of
years into the future. In the case at hand, since the residual value of the land will not be all that much different under either of the alternatives, twenty-five years is not an
unreasonable choice. For a more massive project, longer cut-
off dates might be appropriate. Whatever cut-off date is
chosen, it should be employed consistently across alternatives.

## Δ Town Hall--Daffyland
The corresponding set of figures for the alternative to Paragon Park, Daffyland-by-the-Sea, is shown in Table 4.7. The
basic assumption here, as before, is that if the site is not
used as Daffyland, it will disappear. Therefore, Banks uses
his estimates of the gross municipal public revenues generated
by Daffyland and its tenants and, on the outlay side, the
differences in municipal expenses associated with Daffyland
relative to the site's simply not being there.

Again, Banks has derived his estimates of property taxes
from the assessor's estimates of valuation assessment for
the amusement park and surrounding businesses which would be
displaced by Paragon Park. As mentioned earlier, this involves the amusement park, thirty-three concessions, and
eighteen small commercial endeavors.

As can be expected, the amusement concessionaires contribute relatively large licensing fees and, since the amusement
park is already in existence, these flows begin immediately
at the beginning of our analysis period. Sewerage fees and
sanitation costs are, again, based on the same figures used
for Paragon Park.

The estimated outlays for added police protection are based
on Banks's assumption that due to the nature of the industry,
Daffyland necessitates the hiring of one more full-time police officer and six provisional (summer only) police officers than would be required if the site disappeared. Banks
also estimates that in the beginning, two added full-time
firefighters and two extra call people (volunteers) are
needed to provide adequate fire protection; further, by 1979,
because of the aging of the structures and increasing number
of people using the park, this requirement will double.

The street construction costs are based on the ten dollar
per foot estimate, but the maintenance costs are substan-
tially less, since there is far less pressure from the Daffyland residents to perform all the cosmetic maintenance
necessary on higher-rent residential parkways.

After a little research, Banks finds that the existence
of Daffyland has a negligible effect on both the state revenue redistribution plan and the local school system. This
completes his table and he is ready to make his first comparison.

At a ten percent interest rate, the present value of Table
4.2 is $11,517,300, and that of Table 4.7 is $4,380,130.
Since we are operating from a consistent baseline, this

Table 4.7
Δ Town Hall Under Daffyland Alternative

| | Revenues | | | | Outlays | | | | | Net |
|---|---|---|---|---|---|---|---|---|---|---|
| Year | Property Taxes | Sewerage Fees | Licensing Fees | Change in Cherry Sheet | Police/ Property Protection | Fire Protection | Sanitation Costs | Street Costs | School Costs | Net Cash Flow |
| 1974 | 328,727 | 0 | 13,300 | 0 | -16,906 | -19,748 | 0 | -10,000 | 0 | 295,373 |
| 1975 | 363,599 | 23,600 | 13,300 | 0 | -16,906 | -19,748 | -5,600 | -10,000 | 0 | 320,245 |
| 1976 | 381,038 | 12,600 | 13,300 | 0 | -16,906 | -19,748 | -6,200 | -3,000 | 0 | 360,984 |
| 1977 | 416,883 | 12,600 | 13,300 | 0 | -16,906 | -19,748 | -1,200 | -500 | 0 | 401,329 |
| 1978 | 440,069 | 12,600 | 13,300 | 0 | -16,906 | -19,748 | -1,200 | -500 | 0 | 428,185 |
| 1979 | 458,884 | 12,600 | 13,300 | 0 | -16,906 | -39,596 | -1,200 | -500 | 0 | 427,202 |
| 1980 | 474,684 | 12,600 | 13,300 | 0 | -16,906 | -39,596 | -1,200 | -500 | 0 | 443,002 |
| 1981 | 496,463 | 12,600 | 13,300 | 0 | -16,906 | -39,596 | -1,200 | -1,500 | 0 | 461,467 |
| 1982 | 533,741 | 12,600 | 13,300 | 0 | -16,906 | -39,596 | -1,200 | -1,500 | 0 | 500,745 |
| 1983 | 564,495 | 12,600 | 13,300 | 0 | -16,906 | -39,596 | -1,200 | -500 | 0 | 532,499 |
| 1984 | 591,594 | 12,600 | 13,300 | 0 | -16,906 | -39,596 | -1,200 | -500 | 0 | 559,598 |
| 1985 | 608,811 | 12,600 | 13,300 | 0 | -16,906 | -39,596 | -1,200 | -500 | 0 | 576,815 |
| 1986 | 629,805 | 12,600 | 13,300 | 0 | -16,906 | -39,596 | -1,200 | -500 | 0 | 597,809 |
| 1987 | 650,798 | 12,600 | 13,300 | 0 | -16,906 | -39,596 | -1,200 | -1,500 | 0 | 618,802 |
| 1988 | 667,593 | 12,600 | 13,300 | 0 | -16,906 | -39,596 | -1,200 | -1,500 | 0 | 635,597 |
| 1989 | 688,587 | 12,600 | 13,300 | 0 | -16,906 | -39,596 | -1,200 | -1,500 | 0 | 656,591 |
| 1990 | 705,382 | 12,600 | 13,300 | 0 | -16,906 | -39,596 | -1,200 | -500 | 0 | 673,386 |
| 1991 | 726,375 | 12,600 | 13,300 | 0 | -16,906 | -39,596 | -1,200 | -500 | 0 | 694,379 |
| 1992 | 743,170 | 12,600 | 13,300 | 0 | -16,906 | -39,596 | -1,200 | -500 | 0 | 711,174 |
| 1993 | 764,163 | 12,600 | 13,300 | 0 | -16,906 | -39,596 | -1,200 | -1,500 | 0 | 732,167 |
| 1994 | 780,958 | 12,600 | 13,300 | 0 | -16,906 | -39,596 | -1,200 | -1,500 | 0 | 748,962 |
| 1995 | 797,753 | 12,600 | 13,300 | 0 | -16,906 | -39,596 | -1,200 | -500 | 0 | 765,757 |
| 1996 | 818,746 | 12,600 | 13,300 | 0 | -16,906 | -39,596 | -1,200 | -500 | 0 | 786,750 |
| 1997 | 831,342 | 12,600 | 13,300 | 0 | -16,906 | -39,596 | -1,200 | -500 | 0 | 799,346 |
| 1998 | 852,336 | 12,600 | 13,300 | 0 | -16,906 | -39,596 | -1,200 | -500 | 0 | 820,340 |
| 1999 | 869,131 | 12,600 | 13,300 | 0 | -16,906 | -39,596 | -1,200 | -1,500 | 0 | 837,135 |

implies that the estimate of the change in *Town Hall income* associated with opting for Paragon Park rather than Daffyland is the difference between these two numbers of $7,137,170. This number is not the change in Parable Beach black box income, for we have not yet addressed the non-public accounts, but it is the difference in public income under Banks's assumptions and a ten percent interest rate. At a fifteen percent interest rate, $\Delta$ Town Hall for Paragon Park is about $4,750,000, while that for Daffyland is $2,400,000, for a difference of about $2.3 million.

The Other Accounts
Paragon Park will affect entities other than Town Hall within the black box, as Table 4.1 indicates. Table 4.8 shows the expanded account for the Parable Beach citizens who will be employed on the site under the various hypotheses. As emphasized in Chapter 3, this account should include only the change in take-home pay of these employees between their jobs on the site and what their income would be if the site didn't exist. This difference depends critically on their alternative employment opportunities. Banks knows that at present, 132 of the 170 jobs on the site during the summer months are held by Parablites, while during the winter, forty-seven of the fifty-nine jobs available are filled with locals. On this basis, as well as employment trends in the general area, Banks assumes that the jobs connected with the apartment and marina parcels will contribute twenty percent local hire, while ninety percent of the jobs available in the first fifteen stores to open will go to Parablites. After this, a figure of forty percent local hire is more realistic.

But he has no idea of what the value of the alternative employment opportunities are. Therefore, he decides that the most insightful approach would be to explore the effect different levels of employment would have on his final figures. In order to get a representative view of the full scope of this effect, he chooses to investigate three levels of employment: full employment in the area, a degree of partial unemployment, and complete unemployment. Under conditions of full employment, Banks realizes that whatever wages Parable Beach workers would receive in jobs provided by either alternative would be approximately equal to the wages they would receive elsewhere if neither Paragon Park nor Daffyland existed. As such, neither alternative would contribute anything to Banks's overall analysis. Now, at the other extreme, complete unemployment in the area would mean that any new jobs provided to local inhabitants would provide income to otherwise nonproductive town members. Thus, the total after-tax payrolls of each alternative should be included in the final analysis of the overall impact on the local community.[1]

---

[1]Less any loss in state and federal welfare payments as a result of the increase in employment.

Table 4.8
Parable Beach Employees' Account

| | | Paragon Park | | | | Daffyland | | | |
|---|---|---|---|---|---|---|---|---|---|
| | | | Addition to Local Revenue | | | | Addition to Local Revenue | | |
| Year | Season | Jobs Available | Full Employment | 30% Unemployment | Full Unemployment | Jobs Available | Full Employment | 30% Unemployment | Full Unemployment |
| 1974 | Winter | 0 | 0 | 0 | 0 | 47 | 0 | 49,350 | 164,500 |
| 1975 | Summer | 0 | 0 | 0 | 0 | 132 | 0 | 108,225 | 360,750 |
| 1975 | Winter | 0 | 0 | 0 | 0 | 48 | 0 | 49,560 | 165,200 |
| 1976 | Summer | 0 | 0 | 0 | 0 | 134 | 0 | 109,395 | 364,650 |
| 1976 | Winter | 0 | 0 | 0 | 0 | 48 | 0 | 49,560 | 165,200 |
| 1977 | Summer | 2 | 0 | 2,001 | 6,670 | 136 | 0 | 109,665 | 365,550 |
| 1977 | Winter | 2 | 0 | 5,799 | 19,330 | 49 | 0 | 49,560 | 165,400 |
| 1978 | Summer | 44 | 0 | 6,712 | 22,375 | 139 | 0 | 110,145 | 367,150 |
| 1978 | Winter | 44 | 0 | 9,398 | 31,325 | 50 | 0 | 49,980 | 166,600 |
| 1979 | Summer | 44 | 0 | 6,712 | 22,375 | 142 | 0 | 110,550 | 368,500 |
| 1979 | Winter | 44 | 0 | 9,398 | 31,325 | 50 | 0 | 49,980 | 166,600 |
| 1980 | Summer | 48 | 0 | 9,963 | 33,210 | 145 | 0 | 110,953 | 369,850 |
| 1980 | Winter | 48 | 0 | 13,953 | 46,510 | 52 | 0 | 50,400 | 168,000 |
| 1981 | Summer | 61 | 0 | 11,638 | 38,715 | 148 | 0 | 111,360 | 371,200 |
| 1981 | Winter | 61 | 0 | 16,298 | 54,305 | 53 | 0 | 50,610 | 168,700 |
| 1982 | Summer | 61 | 0 | 11,638 | 38,795 | 151 | 0 | 111,765 | 372,550 |
| 1982 | Winter | 61 | 0 | 16,298 | 54,305 | 54 | 0 | 50,820 | 167,400 |
| 1983 | Summer | 61 | 0 | 11,638 | 38,795 | 154 | 0 | 112,170 | 373,900 |
| 1983 | Winter | 61 | 0 | 16,298 | 54,305 | 54 | 0 | 40,820 | 169,400 |
| 1984 | Summer | 70 | 0 | 13,314 | 44,375 | 157 | 0 | 112,575 | 375,250 |
| 1984 | Winter | 70 | 0 | 18,638 | 62,125 | 55 | 0 | 51,030 | 170,100 |
| 1985 | N/A | 70 | 0 | 31,952 | 106,500 | 212 | 0 | 163,605 | 545,350 |
| 1986 | " | 70 | 0 | 31,952 | 106,500 | 212 | 0 | 163,605 | 545,350 |
| 1987 | " | 70 | 0 | 31,952 | 106,500 | 212 | 0 | 163,605 | 545,350 |
| 1988 | " | 70 | 0 | 31,952 | 106,500 | 212 | 0 | 163,605 | 545,350 |
| 1989 | " | 70 | 0 | 31,952 | 106,500 | 212 | 0 | 163,605 | 545,350 |
| 1990 | " | 70 | 0 | 31,952 | 106,500 | 212 | 0 | 163,605 | 545,350 |
| 1991 | " | 70 | 0 | 31,952 | 106,500 | 212 | 0 | 163,605 | 545,350 |
| 1992 | " | 70 | 0 | 31,952 | 106,500 | 212 | 0 | 163,605 | 545,350 |
| 1993 | " | 70 | 0 | 31,952 | 106,500 | 212 | 0 | 163,605 | 545,350 |
| 1994 | " | 70 | 0 | 31,952 | 106,500 | 212 | 0 | 163,605 | 545,350 |
| 1995 | " | 70 | 0 | 31,952 | 106,500 | 212 | 0 | 163,605 | 545,350 |
| 1996 | " | 70 | 0 | 31,952 | 106,500 | 212 | 0 | 163,605 | 545,350 |
| 1997 | " | 70 | 0 | 31,952 | 106,500 | 212 | 0 | 163,605 | 545,350 |
| 1998 | " | 70 | 0 | 31,952 | 106,500 | 212 | 0 | 163,605 | 545,350 |
| 1999 | " | 70 | 0 | 31,952 | 106,500 | 212 | 0 | 163,605 | 545,350 |

For the conditions of partial unemployment, Banks decides
to assume that the employees on either project would average
thirty percent more working on the site than if the site
disappeared. Using this assumption and recent data on local
salaries, Banks is able to arrive at the figures shown in
Tables 4.9 and 4.10. As could be expected, the more
employment-intensive Daffyland contributes more to this account.

In considering the account for the construction labor employed by the alternative developments, Banks again has to
arrive at some reasonable assumptions concerning alternative
employment opportunities. In addition, he must estimate how
many Parablites are employed in the construction industry
and are fluid enough in their jobs to work in the locale of
their choice. He decides on the basis of recent planning
council figures that employment will run as shown in Table
4.9 on the Paragon Park construction site.

Again, Banks faces the problem of alternative employment
and again he chooses to perform the analysis for the same
three hypotheses used earlier. Since Daffyland already
exists, there is no contribution from construction on its
side of the ledger. The revenue additions in the case of the
Paragon Park development are shown in Table 4.10.

In relation to concessionaires--the persons owning and
operating commercial establishments in each of Banks's two
alternative uses--Table 4.11 is the summary of his findings.
For this account, Banks must assume how many of the concessionaires are Parablites. Changes in income to non-Parablites
are ignored. Further, he must assume how much the Parablite
concessionaires would make elsewhere if the site disappeared.

These figures again necessitate a number of assumptions
on Banks's part. First, research he conducts into the economic base in the area shows that with the Daffyland alternative, fifteen of the eighteen businesses stayed open year-
round, while only three of the thirty-three concessions were
open a full twelve months. Using a five-month summer season
and the assumptions shown below, Banks is able to arrive at
his entries for this case. Those for Paragon Park are made
using the openings of the businesses scheduled as in Banks's
original timetable.

## The Parable Beach Customers' Account

The next account in Table 4.1 covers changes in real income
which would result from market price changes to Parable Beach
customers of the establishments in the alternative development. For example, if one development included a hardware
store which priced its products lower than those in the hardware store of the other development, this would be an increase
in real income to the Parable Beach customers of the lower-
priced store. However, Banks sees no reason to believe the
prices will be very different and comes to the conclusion
that the Δ Parable Beach customers' account will be negligible in both cases. [2]

--------------------------------------------------------------

[2] A change in real customer income could also arise if one

Table 4.9
Employment Estimates for Construction of Paragon Park

| Year | Project | Number of Employees | | | |
| --- | --- | --- | --- | --- | --- |
| | | State | | Local | |
| | | Skilled | Unskilled | Skilled | Unskilled |
| 1974 | Clearing | 4 | 12 | 0 | 5 |
| 1975 | Construction, 304 apartments | 60 | 40 | 2 | 20 |
| 1976 | Construction, 304 apartments | 40 | 15 | 2 | 8 |
| | 15 stores | 30 | 20 | 1 | 10 |
| | Marina | 30 | 10 | 5 | 5 |
| 1977 | Construction, 15 stores | 30 | 10 | 1 | 5 |
| | Marina | 30 | 10 | 1 | 5 |
| 1978 | Construction, 10 stores | 30 | 10 | 1 | 5 |
| | 400 apartments | 60 | 30 | 3 | 15 |
| 1979 | Construction, 10 stores | 30 | 30 | 1 | 5 |
| | 400 apartments | 50 | 20 | 1 | 10 |
| 1980 | Construction, 400 apartments | 40 | 10 | 1 | 5 |
| 1981 | | 0 | 0 | 0 | 0 |
| 1982 | Clearing | 4 | 12 | 0 | 6 |
| 1983 | Construction, 5 stores | 20 | 10 | 1 | 5 |
| | Motel | 30 | 15 | 1 | 8 |

Table 4.10
Δ Construction Labor on Paragon Park

|       | Addition to Revenues | | |
|-------|-----------------|------------------|--------------------|
| Year  | Full Employment | 30% Unemployment | Full Unemployment  |
| 1974  | --              | $ 5,940          | $ 19,800           |
| 1975  | --              | 24,300           | 81,000             |
| 1976  | --              | 40,770           | 135,900            |
| 1977  | --              | 14,400           | 48,000             |
| 1978  | --              | 28,800           | 96,000             |
| 1979  | --              | 19,350           | 64,500             |
| 1980  | --              | 7,200            | 24,000             |
| 1981  | --              | --               | --                 |
| 1982  | --              | 5,940            | 17,800             |
| 1983  | --              | 17,270           | 57,900             |
| 1984  | --              | --               | --                 |

The Parable Beach Apartment Tenants' Account
Some of the Paragon Park apartments will undoubtedly be occu-
pied by Parablites who will feel they are at least as well
off as before as a result of the opportunity to rent at Paragon Park.  If they consider themselves much better off--for
example, if a family would have been willing to pay $500 for
an apartment for which they actually have to pay $250--this
is an increase in real municipal income.  However, Banks
knows the apartments will be competitively priced, with perhaps a premium for the view, which means it is unlikely that
many such Parablites exist, for they could have purchased
equivalent space elsewhere.  Therefore, he puts the Δ Parable
Beach tenants' account down as negligible.

The Parable Beach Boatowners' Account
Parablites who choose to moor their boats at the proposed
marina may experience real income changes.  However, since
the local price of mooring space is likely to change drasti-
cally with the provision of the marina, the analysis of
these income changes requires the introduction of a couple
of new concepts.  Banks decides to put off the analysis of
this account for the moment, and to act as if it were

development included a hardware store and the other did not,
necessitating a longer trip for hardware purchases on the
part of Parablites if the local store did not increase its
price to the point where Parablites were indifferent to the
longer trips.  Banks, a long-time resident, has noted that
local stores invariably follow the latter course.

Table 4.11
Δ Concessionaires

| Year | Development Owners Out-of-State | | Development Owners State Residents | |
|------|------------------|-----------|------------------|-----------|
| | Paragon Park | Daffyland | Paragon Park | Daffyland |
| 1974 | 0 | 41,400 | 0 | 41,400 |
| 1975 | 0 | 197,800 | 0 | 197,800 |
| 1976 | 0 | 197,800 | 0 | 197,800 |
| 1977 | 0 | 197,800 | 0 | 197,800 |
| 1978 | 45,000 | 197,800 | 45,000 | 197,800 |
| 1979 | 60,000 | 197,800 | 60,000 | 197,800 |
| 1980 | 90,000 | 197,800 | 90,000 | 197,800 |
| 1981 | 150,000 | 197,800 | 150,000 | 197,800 |
| 1982 | 165,000 | 197,800 | 165,000 | 197,800 |
| 1983 | 180,000 | 197,800 | 180,000 | 197,800 |
| 1984 | 210,000 | 197,800 | 210,000 | 197,800 |
| 1985 | 210,000 | 197,800 | 210,000 | 197,800 |
| 1986 | 210,000 | 197,800 | 210,000 | 197,800 |
| 1987 | 210,000 | 197,800 | 210,000 | 197,800 |
| 1988 | 210,000 | 197,800 | 210,000 | 197,800 |
| 1989 | 210,000 | 197,800 | 210,000 | 197,800 |
| 1990 | 210,000 | 197,800 | 210,000 | 197,800 |
| 1991 | 210,000 | 197,800 | 210,000 | 197,800 |
| 1992 | 210,000 | 197,800 | 210,000 | 197,800 |
| 1993 | 210,000 | 197,800 | 210,000 | 197,800 |
| 1994 | 210,000 | 197,800 | 210,000 | 197,800 |
| 1995 | 210,000 | 197,800 | 210,000 | 197,800 |
| 1996 | 210,000 | 197,800 | 210,000 | 197,800 |
| 1997 | 210,000 | 197,800 | 210,000 | 197,800 |
| 1998 | 210,000 | 197,800 | 210,000 | 197,800 |
| 1999 | 210,000 | 197,800 | 210,000 | 197,800 |

| | Business | Concessions |
|---|----------|-------------|
| Net summer earnings | $900/month | $1000/month |
| Net winter earnings | $1300/month | $400/month |

negligible. He makes a note to himself to later go back and
check this assumption. (See the appendix to this chapter.)

## Putting It All Together

Banks also decides to ignore for the moment the change in
Parable Beach real income associated with the respending of
the direct effects he has estimated. In a moment, we will
see why.

Ignoring for the moment, then, the boatowners and respending, Banks is ready to put all his estimates together.
Table 4.12 shows the total yearly change in Parable Beach
income for each alternative for each of the three employment
hypotheses relative to the site's disappearing. These figures are simply the sum of all the accounts for each year
for each alternative.

Table 4.13 shows the present value of each column in Table
4.12 for a range of interest rates, and Figures 4.1 a, b, and
c plot these present values for each employment hypothesis.
The vertical difference between the solid lines (Paragon
Park) and the dotted lines (Daffyland) in these figures are
Banks's estimate of the present value change in Parable Beach
income associated with opting for Paragon Park rather than
Daffyland for the corresponding employment hypothesis and
interest rate before consideration of the boatowners and respending. This change ranges from a high of +$12.6 million
(low interest rate, full employment) to a low of -$1.1 mil-
lion (high interest rate, complete unemployment). "Plus"
here means town has higher real income with Paragon Park;
"minus" indicates higher real income with Daffyland. The
figures show the break-even points. The lower the interest
rate and the lower the unemployment, the more Paragon Park
is favored. Notice, however, that the differences are no-
where near as great as some of the figures which the prodevelopment and antidevelopment forces have been throwing
around. For intermediate assumptions about relevant interest
rates and unemployment, the difference is in the neighborhood of about $1 million, which corresponds to handing every
woman, man, and child in Parable Beach $100 worth of income
on a one-shot basis or about twelve dollars per year over
the twenty-five year analysis period. Under Banks's assumptions, these are the magnitudes of the income changes with
which Parable Beach must be concerned in deciding between
the alternative allocations of the sites. Depending on the
town's evaluation of the differences in the non-market effects associated with the alternatives, the environmental
and aesthetic values, the town might quite rationally decide
to forgo such increases in real income and opt for Daffyland.

In any event, it will be basing its decision on a coherent
assessment of the change in municipal income rather than on
a potpourri of developer and anti-development figures, most
of which are only remotely related to this change.

Table 4.12
Cash Flow Table for Parable Beach Black Box

Cash Flows (Local Analysis)

| Year | Paragon Park | | | Daffyland | | |
|---|---|---|---|---|---|---|
| | Full Employment | 30% Unemployment | Full Unemployment | Full Employment | 30% Unemployment | Full Unemployment |
| 1974 | 5,940 | 10,930 | 25,790 | 336,773 | 386,123 | 501,273 |
| 1975 | 177,244 | 201,544 | 258,244 | 548,045 | 705,830 | 1,073,995 |
| 1976 | 393,699 | 434,469 | 529,599 | 558,784 | 717,739 | 1,088,634 |
| 1977 | 626,407 | 648,607 | 700,407 | 559,129 | 758,354 | 1,130,579 |
| 1978 | 1,047,812 | 1,092,722 | 1,197,512 | 621,985 | 782,110 | 1,155,735 |
| 1979 | 1,175,631 | 1,211,091 | 1,293,831 | 625,002 | 785,532 | 1,160,102 |
| 1980 | 1,600,664 | 1,631,770 | 1,704,384 | 640,802 | 802,157 | 1,178,652 |
| 1981 | 1,715,683 | 1,743,619 | 1,808,783 | 659,267 | 821,237 | 1,199,187 |
| 1982 | 1,809,084 | 1,842,960 | 1,921,984 | 698,545 | 861,130 | 1,240,495 |
| 1983 | 1,957,812 | 2,003,018 | 2,108,812 | 730,299 | 873,289 | 1,253,599 |
| 1984 | 2,071,615 | 2,103,567 | 2,178,115 | 757,398 | 921,003 | 1,402,748 |
| 1985 | 2,124,668 | 2,156,620 | 2,231,168 | 774,615 | 938,220 | 1,319,965 |
| 1986 | 2,213,091 | 2,245,043 | 2,319,591 | 795,609 | 959,214 | 1,340,659 |
| 1987 | 2,283,829 | 2,315,781 | 2,390,329 | 816,602 | 980,207 | 1,361,952 |
| 1988 | 2,336,882 | 2,368,834 | 2,443,382 | 833,397 | 997,002 | 1,378,747 |
| 1989 | 2,425,305 | 2,457,257 | 2,531,805 | 854,391 | 1,017,996 | 1,399,741 |
| 1990 | 2,478,358 | 2,510,310 | 2,584,858 | 871,186 | 1,034,791 | 1,416,536 |
| 1991 | 2,549,096 | 2,581,048 | 2,655,596 | 892,179 | 1,055,784 | 1,437,529 |
| 1992 | 2,602,150 | 2,634,102 | 2,708,650 | 908,974 | 1,072,579 | 1,454,324 |
| 1993 | 2,672,888 | 2,704,840 | 2,779,388 | 829,967 | 1,093,572 | 1,475,317 |
| 1994 | 2,743,626 | 2,775,578 | 2,850,126 | 846,762 | 1,110,367 | 1,492,112 |
| 1995 | 2,814,364 | 2,846,316 | 2,920,864 | 863,557 | 1,127,162 | 1,508,907 |
| 1996 | 2,867,418 | 2,899,370 | 2,973,918 | 884,550 | 1,148,155 | 1,529,900 |
| 1997 | 2,938,156 | 2,970,108 | 3,044,656 | 897,146 | 1,160,751 | 1,542,496 |
| 1998 | 3,008,894 | 3,040,846 | 3,115,394 | 1,018,140 | 1,181,745 | 1,563,490 |
| 1999 | 3,061,947 | 3,093,899 | 3,168,447 | 1,034,935 | 1,198,540 | 1,580,285 |

Table 4.13
Net Present Values for Alternative Developments: Local (Parable Beach) Analysis
(Millions of 1974 Dollars)

| Interest Rate (%) | Paragon Park | | | Daffyland | | |
|---|---|---|---|---|---|---|
| | Full Employment | 30% Unemployment | Full Unemployment | Full Employment | 30% Unemployment | Full Unemployment |
| 6 | 22.2356 | 22.6602 | 23.6170 | 9.6320 | 11.9214 | 16.9214 |
| 10 | 13.9873 | 14.2814 | 14.9720 | 6.6694 | 8.2632 | 11.8368 |
| 20 | 5.7333 | 5.8898 | 6.2896 | 3.5312 | 4.3870 | 6.3722 |
| 40 | 1.9149 | 1.9995 | 2.1999 | 1.9819 | 2.2840 | 3.3333 |

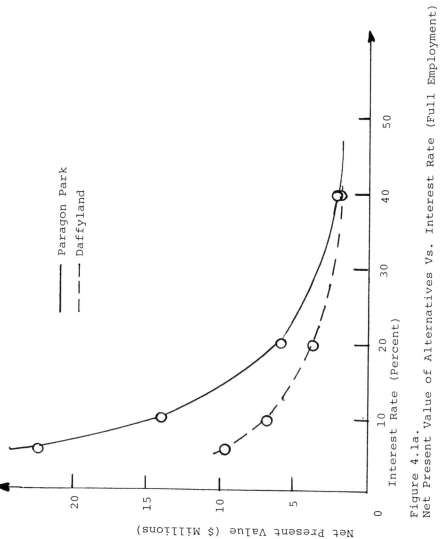

Figure 4.1a.
Net Present Value of Alternatives Vs. Interest Rate (Full Employment)

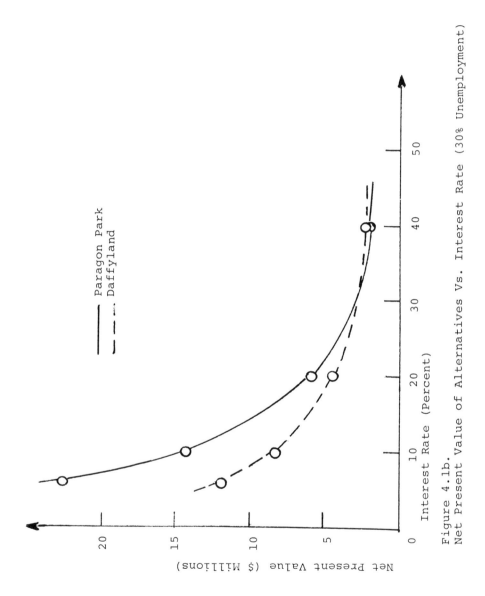

Figure 4.1b.
Net Present Value of Alternatives Vs. Interest Rate (30% Unemployment)

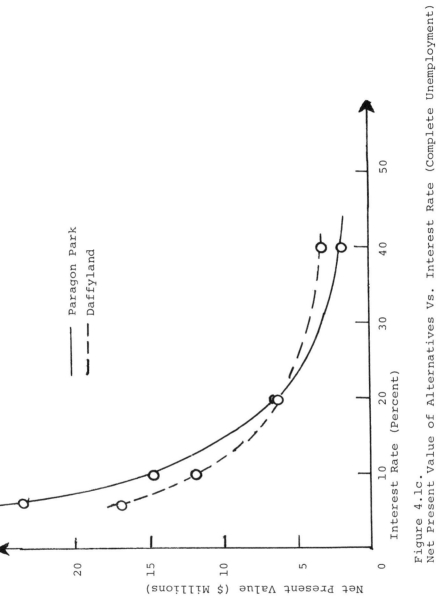

Figure 4.1c.
Net Present Value of Alternatives Vs. Interest Rate (Complete Unemployment)

## The Effect of Respending

Up to this point, Banks has ignored the effect of respending of the direct changes on real income associated with the two alternatives.

The impact of respending on real black-box income depends firstly on the percentage of the changes in direct real income which are spent within the black box and secondly, on the degree of unemployment or underemployment in the markets where this respending takes place. Thus, the smaller the black box and the smaller the degree of unemployment, the smaller the impact of respending.

Notice also that if one is willing to assume that the pattern of respending of all the direct changes in income for each alternative is the same--that is, the percentage of local respending of an increase in Town Hall income is similar to the percentage of local spending of, say, employee income--then *whatever the net multiplier is for a given employment hypothesis, it applies to all the direct increase in municipal income equally.* (Despite this fact, in the public coastal zone debate, people often apply multipliers only to increases in local wages.) This implies, for a given employment hypothesis, we might as well apply the net multiplier to the total difference between the alternatives. In short, for a given employment hypothesis, *the net multiplier, whatever its value, will not change the ranking of alternatives* as long as the pattern of local expenditures resulting from the various direct changes in income is roughly similar for all alternatives. Under this assumption, the multiplier is completely irrelevant to rankings.

Rather, the effect of the multiplier will be to favor the alternative which offers the higher black-box income under conditions of higher unemployment. To see this, let's suppose the increase in local income as a percent of expenditures is zero per round under full employment, twenty-five percent per round under the moderate unemployment hypothesis, and fifty percent per round under the complete unemployment hypothesis. Applying the resulting multipliers to our earlier differences, we arrive at Table 4.14.

From Table 4.14, we see the overall net effect of respending is to somewhat increase the advantage of the alternative which is favored by increasing unemployment, in this case Daffyland. Thus, if we are unsure of what the future employment situation is going to be, respending effects will tend to lead us a little more toward the employment-intensive alternative. This is the germ of truth underlying the use of the multiplier concept in resource allocation problems. Needless to say, in most of the discussion surrounding potential developments, the net effect of respending is much overrated.

The problem still remains. How large is the net multiplier? Unfortunately, this is not easily answered. One can put bounds on it. Under full employment it is zero and under complete and utter unemployment it approaches the percentage value added in the local market multiplied by the

Table 4.14
Impact of Respending Assuming a Ten Percent Interest Rate
(Millions of 1974 Dollars)

| Condition | Element | Paragon Park | Daffyland |
|---|---|---|---|
| Full Employment | $\Delta$ Direct | 13.9873 | 6.6694 |
| | Multiplier | 0 | 0 |
| | $\Delta$ Respending | 0 | 0 |
| | $\Delta$ Total | 13.9873 | 6.6694 |
| | Differential | 7.3179 | |
| Intermediate Unemployment | $\Delta$ Direct | 14.2814 | 8.2632 |
| | Multiplier | 1.33 | 1.33 |
| | $\Delta$ Respending | 4.7129 | 2.7268 |
| | $\Delta$ Total | 18.9943 | 10.9900 |
| | Differential | 8.9943 | |
| Full Unemployment | $\Delta$ Direct | 14.9720 | 11.8368 |
| | Multiplier | 2.00 | 2.00 |
| | $\Delta$ Respending | 14.9720 | 11.8368 |
| | $\Delta$ Total | 29.9440 | 23.6736 |
| | Differential | 6.2704 | |

percentage of direct increases in black-box income which are
respent locally. In those rare cases where the respending
effects will be critical to the allocative decisions, the
analyst has little choice but to actually trace them down.
Fortunately, due to the very rapid fall-off, one almost never
has to trace these effects more than one or two rounds.

## THE TREATMENT OF PRICE/QUANTITY CHANGES

### The Parable Beach Boatowners' Account

Banks decided, in his first cut at the problem, to ignore
any changes in the real income of the Parable Beach residents
who would make use of the Paragon Park marina. He now decides to go back and actually work out this account. As we
shall see, this analysis will allow us to illustrate several
interesting points.

The proposed marina will handle 500 boats. At present,
dock space in the Parable Beach area is quite limited--no
more than fifty boats. As a result, the available dock space
commands premium rates: twenty dollars per boat-foot for
the summer six months. Only ten Parablites keep their boats
on this expensive space. Several hundred Parablites own
boats, almost all of which are presently on moorings for
which they pay fifty dollars per season to the town harbor-
master.

The developers estimate that in order to fill the marina,
the rate will have to drop to ten dollars per boat-foot, at
which point they can still make a handsome profit. Banks
estimates that at ten dollars per foot, 100 Parablites,
whose boats average thirty feet in length, will abandon
their moorings and take space at the marina, paying $30,000
a year for the privilege. The question, then, is: does
this outlay represent a change in Parable Beach real income?

Whenever a significant change in either the price or
quantity of a good is implied by one of the alternatives
under analysis, it's a good idea to sketch the demand curve
for that good or service. This has been done for the marina
case in Figure A4.1.

In this figure, we have assumed that the demand for boat
space is inversely proportional to price. Under this assumption, the single-hatched area is the increase in real income
of the boatowners who opt to rent at the marina. To see
this, consider a boatowner who is at point x on the demand
curve. He was willing to pay as much as fifteen dollars
per foot for boat space, but no more. Thus, ante marina, at
twenty dollars per foot, he didn't purchase space. He now
has the opportunity to buy space at ten dollars per foot.
In terms of his own willingness to pay, he is five dollars
per foot ahead on net as a result of the construction of
the marina and the subsequent price reduction. Similarly,
the man at y who was almost willing to pay twenty dollars
per foot is close to ten dollars per foot ahead, while the
man at z who is just barely willing to pay ten dollars per
foot essentially comes out even. He gets his boat on a
dock but the value he places on the other goods and services
he would have consumed if he didn't pay the dockage rates is

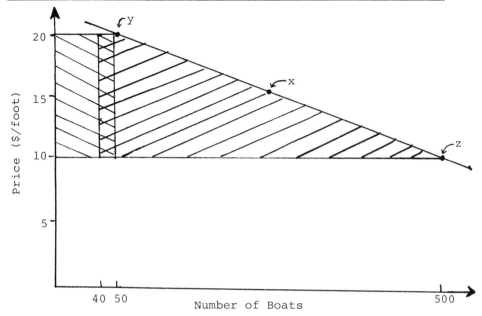

Figure A4.1.
Demand curve for dock space.

such that he is only barely better off in terms of his own
willingness to pay.
    The sum of the differences between the maximum of what
the boatowners would be willing to pay and what they actually
end up paying, the single-hatched area, is the total increase
in income to all the boat owners at the marina.  Under our
assumptions, this number, annually, is

$10 x 30 feet x 500/2 = $75,000 .

Now 100 of these boatowners are Parablites, so a portion of
this increase in real income accrues to the black box.  Banks
knows the Parablite boat owner is on the average poorer than
the outside boat owner, so he assumes the 100 Parablite
owners are concentrated in the rightmost portion of the demand curve, and estimates the Parablites' share of this increase in real wealth at fifteen percent of the $75,000, or
$9,750, two-thirds of what it would be if the Parablites
were spread evenly over the demand curve.
    The double-hatched area is the increase in real income to
Parablites who are presently purchasing mooring space at
twenty dollars per foot and who, after the price drop, will
have it at ten dollars per foot, a clear increase in their
real income.  However, whether this is a net increase in
real municipal income depends on whether or not the original
fifty spaces were *owned* by a Parablite or not.  If not, then

it is an increase.  If yes, then the double-hatched area is
a wash as far as Parable Beach is concerned--the increase in
real income of the Parable Beach boatowners is matched by a
decrease in the income of the old marina owner, and further,
the dotted area, $10 x 30 x 40 - $1,200, is a net loss, for
outside boatowners gain and a Parable Beach citizen loses.
In the case at hand, Banks knows the old marina is owned by
his father-in-law, and reluctantly takes $1,200 annually off
the $\Delta$ Parable Beach marina tenants' account.[1]  The net annual Parable Beach income due to the marina's output, then,
is $8,750 per year, starting four years hence, when the ma-
rina is completed.  The present value, then, over the twenty-
five year analysis period, is

$$\sum_{n=3}^{25} \frac{8,750}{(1 + i)^n}$$

which is $111,854 at six percent and $19,443 at forty-five
percent.  This amount should be added to the other changes
in municipal income associated with the Paragon Park high-
rise development.

--------------------------------------------------------------

[1]Also, if the town moorings are not re-employed, then there
is an annual loss in Town Hall income of 100 x 50, less the
expenses associated with maintaining the moorings.  Banks has
assumed they will be re-employed at the old price.

## Chapter 5

THE ORONOCO REFINERY PROPOSAL

## Once More Into the Breach

It is a Monday night six months later. A subdued Parable Beach town council is meeting in closed session. A weary George Banks is also present, fiddling with a balky transparency projector. It has been a difficult six months for the council. Their decision on Paragon Park was not a universally popular one. On top of this, they have just completed a grueling, extremely heated series of public hearings on a much more momentous decision facing the town.

The Oronoco Oil Company has approached the town with a proposal to place a 250,000 barrel per day refinery in the abandoned sand and gravel pit area at the root of the Parable Beach peninsula. At present this 650-acre site is owned by Ephraim Ahab, a one-time resident of the town who has lived in Florida since the gravel pit operations shut down twenty years ago. This land is presently unused except for one corner which contains the town dump, for which use the town pays Ahab an amount equal to his property tax.

The plan involves placing a single-buoy mooring three miles off South Beach through which crude oil will be unloaded and piped to shore. The refined products produced by the installation will be distributed by truck and pipeline to the greater Schrod City market. The bulk of the site will be taken up by crude, intermediate, and products storage tanks totalling over twelve million barrels. The actual process units will occupy less than ten percent of the site, but several of the units will be close to 150 feet tall, and the entire complex will be dominated by a 300 foot stack. The project will also involve a ten-acre wastewater treatment lagoon, the consumptive use of three million gallons of fresh water per day, and the construction of a wastewater outfall just south of South Beach. The debate within the town, the entire state, and the whole region concerning this project has reached a fever pitch. Within the town, people are talking of little else. It has been the cause of domestic quarrels and barroom brawls. An advisory referendum split sixty-forty in favor of the refinery, but the summer residents are crying foul since they were not represented in the voting, which was open only to voters registered in the town. The town council is stuck at the center of this debate.

Simon Montfort calls the meeting to order. He begins, "As you all know, the purpose of this meeting is to receive George Banks's analysis of the effect of the refinery on the town's real income. [Montfort is beginning to learn the jargon.] For tonight, I suggest we concentrate exclusively on the economic aspects of the refinery and wait until Wednesday to begin discussing the environmental

aspects.  At Wednesday's meeting we will receive the report
of Professors Vachs and Wayne on the atmospheric and marine
pollutants associated with the proposal.  George, would you
begin please?"

George Banks's Report
"Thank you, Simon.  As you know, I have been asked to subject the developer's economic figures to the same sort of
analysis used in the Paragon Park case to estimate the
change in Parable Beach real income associated with accepting the Oronoco refinery in the town.  In order to do this,
I have talked at length with the developers and, at town
expense, have visited several independent refinery experts.
    "The Oronoco proposal envisions a rather straightforward
fuels refinery, consuming about 250,000 barrels of Persian
Gulf crude per day.  The general characteristics of this
plant are shown in Slide 1.
    "Construction would require a six-month detailed design
period and two years of site preparation and erection.  The
construction work crew would be approximately 2,200 people
for the first twelve months and 1,000 people for the second
twelve months.  Total on-site construction payroll will be
about $40 million out of a $400 million total investment on
the part of the developer.  The permanent work force will be
approximately 350 people with an annual payroll of about
$3.5 million.
    "In order to investigate the effect of this proposal on
Parable Beach real income, I have set up the accounts shown
in Slide 2.  For this analysis, I have chosen as my baseline
a projection of the status quo.  That is, all my computa-
tions will be relative to continuing to use the area for a
dump.  *I have also made the very important assumption that
there is no other possible use of the site--it will either
be a refinery or a dump*.  After completing the analysis under this assumption, we will want to return and re-examine
this hypothesis.
    "Let us begin with the town hall account.  As you know,
in our state, towns are only allowed to tax on the basis of
real property.  What is and what is not real property is a
legal question and we can expect litigation and a judicial
decision on the matter.  However, discussions with a number
of tax experts and a review of similar litigation indicates
to me that we will not be able to assess the refinery for
more than about $70 million, something less than twenty per-
cent of its construction cost.  Assuming this to be the
case, we must ask ourselves what the tax rate will be in
the future, which of course depends on how much the town
wishes to spend in municipal services.  With respect to this
variable, I have studied two extremes:

    A.  The town decides that each taxpayer will pay the same
rate with the refinery as without.  In this case, any net
impact of the refinery on town hall income will go to increasing the quality and quantity of municipal services.

George Banks's Slide 1
General Characteristics of Oronoco Proposal

---

Crude
250,000 barrels per day (bpd) of medium sulphur, light
Arabian

Products
 7,000 bpd of LPG
20,000 bpd of lead-free, premium gasoline
30,000 bpd of lead-free, regular gasoline
80,000 bpd of No. 2 fuel oil, .1% sulphur
96,000 bpd of No. 5 fuel oil, .3% sulphur

Refinery fuel
13,000 bpd refinery fuel oil
 4,000 bpd equivalent refinery gas

Sulphur
480 metric tons/day

Storage
Crude          50,000,000  (10 tanks)
Intermediate   25,000,000  (50 tanks)
Oil products   50,000,000  (18 tanks--6 heated)
Gas products      200,000  (20 tanks)
               _____
              125,000,000

---

George Banks's Slide 2
Accounts for Municipal Analysis of Refinery (Baseline is
Status Quo)

---

Δ Town Hall

Δ Parable Beach construction labor
Δ Parable Beach permanent employees

Δ Parable Beach landowners
Δ Parable Beach oil consumers

Δ Responding

---

B.  The town decides it wishes to provide the same level
of public municipal services with the refinery as without.
In this case, any net impact of the refinery on town hall
income will be transferred to the town's property owners in
the form of lower taxes.

"Let's start with Alternative A (Slide 3).  In this case,
the property tax revenue calculation is quite simple.  We are
presently taxing ourselves at the rate of $100 per thousand-
dollar valuation and this rate has recently been increasing
at one and one-half percent per year in real terms, that is,
after correction for inflation.  This, together with the two
and one-half year construction period, yields the revenues
shown in the leftmost column in Slide 3.
"On the outlay side, the refinery will involve additional
town expenses for:
1.  road construction and maintenance;
2.  sewage construction and maintenance;
3.  replacing the town dump;
4.  police protection and environmental monitoring;
5.  fire protection;
6.  cleaning, painting public facilities.
"With respect to Item 1, the town is fortunate in that
the refinery is located near the southwestern border of the
town, whence all the construction materials and most of the
workers will come.  This means that the neighboring town of
Horseham will bear the brunt of this traffic.[1]
"However, the refinery will require the construction of
one mile of four-lane road within the town and approximately
four miles of existing town road will be subjected to about
5,100 additional trips per day by construction workers,
equipment, and material during the construction period.  The
refinery's operations will imply about 600 vehicle trips per
day and, since about twenty percent of the refinery output
will be delivered direct by truck, about 1,000 truck trips
daily.  This represents an increase in usage of fifty percent on these roads during the construction period and of
fifteen percent during the operating period, much of it
heavy truck traffic.  My discussion with the town road
people indicates the overall effect will be to increase maintenance costs on these roads about fifty percent over what
they would be without the refinery.  This amounts to a little
over $134,000 a year in addition to the initial $470,000 construction bill.
"Sewage disposal will have both a positive and a negative
effect on this account.  On the plus side, the town will be
able to assess the refinery the standard sewage fees; on
the negative side, the town will be required to construct
additional conduits and conduit access and accept an increase
------------------------------------------------------------
[1]Horseham, a wealthy suburban community, has been adamant in
its opposition to the refinery, and is attempting to mobi-
lize forces at the state level to stop the project.

George Banks's Slide 3
Δ Town Hall--Oronoco Oil Refinery Assuming No Decrease in Property Taxes (Millions of 1974 Dollars)

| Year | Property Taxes | Cleaning and Painting | Sewer Fees | Road Construction and Maintenance | Sewer Construction and Maintenance | Replacing Town Dump | Police Protection* | Fire Protection | Net Cash Flow |
|---|---|---|---|---|---|---|---|---|---|
| 1974 | 0 | 0 | 0 | 0 | 0 | 0 | 0 | 0 | 0 |
| 1975 | 2.141 | 0 | 0.0255 | -0.0678 | -0.0250 | -0.1350 | -0.0598 | -0.0950 | 1.7639 |
| 1976 | 5.148 | 0 | 0.0255 | -0.8847 | -0.0250 | -0.1364 | -0.0678 | -0.0700 | 3.5039 |
| 1977 | 5.936 | -0.015 | 0.0120 | -0.1348 | -0.0024 | -0.1378 | -0.0611 | -0.1341 | 5.4638 |
| 1978 | 7.560 | -0.015 | 0.0120 | -0.1347 | -0.0024 | -0.1392 | -0.0611 | -0.1341 | 7.0863 |
| 1979 | 7.700 | -0.015 | 0.0120 | -0.1348 | -0.0024 | -0.1406 | -0.0611 | -0.1341 | 7.2250 |
| 1980 | 7.840 | -0.015 | 0.0120 | -0.1347 | -0.0024 | -0.1420 | -0.0611 | -0.1341 | 7.3635 |
| 1981 | 7.980 | -0.015 | 0.0120 | -0.1348 | -0.0024 | -0.1434 | -0.0611 | -0.1341 | 7.5022 |
| 1982 | 8.120 | -0.015 | 0.0120 | -0.1348 | -0.0024 | -0.1448 | -0.0611 | -0.1341 | 7.6415 |
| 1983 | 8.260 | -0.015 | 0.0120 | -0.1347 | -0.0024 | -0.1462 | -0.0611 | -0.1341 | 7.7794 |
| 1984 | 8.400 | -0.015 | 0.0120 | -0.1348 | -0.0024 | -0.0477 | -0.0611 | -0.1341 | 7.9178 |
| 1985 | 8.540 | -0.015 | 0.0120 | -0.1347 | -0.0024 | -0.1492 | -0.0611 | -0.1341 | 8.0564 |
| 1986 | 8.680 | -0.015 | 0.0120 | -0.1348 | -0.0024 | -0.1507 | -0.0611 | -0.1341 | 8.1948 |
| 1987 | 8/820 | -0.015 | 0.0120 | -0.1347 | -0.0024 | -0.1522 | -0.0611 | -0.1341 | 8.3506 |
| 1988 | 8.960 | -0.015 | 0.0120 | -0.1348 | -0.0024 | -0.1537 | -0.0611 | -0.1341 | 8.3816 |
| 1989 | 9.100 | -0.015 | 0.0120 | -0.1347 | -0.0024 | -0.1552 | -0.9611 | -0.1341 | 8.6204 |
| 1990 | 9.240 | -0.015 | 0.0120 | -0.1348 | -0.0024 | -0.1568 | -0.9611 | -0.1341 | 8.7587 |
| 1991 | 9.380 | -0.015 | 0.0120 | -0.1347 | -0.0024 | -0.1584 | -0.9611 | -0.1341 | 8.8972 |
| 1992 | 9.520 | -0.015 | 0.0120 | -0.1348 | -0.0024 | -0.1600 | -0.9611 | -0.1341 | 9.0555 |
| 1993 | 9.660 | -0.015 | 0.0120 | -0.1347 | -0.0024 | -0.1616 | -0.9611 | -0.1341 | 9.1740 |
| 1994 | 9.800 | -0.015 | 0.0120 | -0.1348 | -0.0024 | -0.1632 | -0.9611 | -0.1341 | 9.3123 |
| 1995 | 9.940 | -0.015 | 0.0120 | -0.1348 | -0.0024 | -0.1648 | -0.9611 | -0.1341 | 9.4508 |
| 1996 | 10.080 | -0.015 | 0.0120 | -0.1348 | -0.0024 | -0.1664 | -0.9611 | -0.1341 | 9.5831 |
| 1997 | 10.220 | -0.015 | 0.0120 | -0.1347 | -0.0024 | -0.1681 | -0.9611 | -0.1341 | 9.6675 |
| 1998 | 10.360 | -0.015 | 0.0120 | -0.1348 | -0.0024 | -0.1698 | -0.9611 | -0.1341 | -9.8657 |
| 1999 | 10.500 | -0.015 | 0.0120 | -0.1347 | -0.0024 | -0.1715 | -0.9611 | -0.1341 | 10.0041 |

*This includes increased environmental quality control procedures.

in operation and maintenance expenses which will accrue from
refinery tie-in.

"The refinery will also require an alternative means of
disposing of the town's trash.  At present the town generates waste at the rate of five pounds per person per day,
or about twenty-five tons per day.  Assuming the pits are
no longer available to us, the only alternative in the near
term involves trucking to a sanitary landfill twenty miles
to the southwest.  Conversations with local contractors indicate that this will cost fifteen dollars per ton more than
our present collection and dumping costs, or about $135,000
per year.

"Discussion with the police chief indicates he would like
one additional permanent police officer and two additional
provisional police officers.  These people would be primarily employed in handling the additional traffic generated
by the refinery.  In this column, I have also included money
for an environmental quality control expert and an assistant
along with rudimentary lab and support facilities on the
assumption that the town will opt to run its own environmental monitoring program in addition to those conducted by
state and federal agencies.  In the fire protection column,
I have allowed for the addition of a complete pumping unit
including one new officer, eight permanent firefighters and
eight new callmen, one new firehouse with associated equipment, and a new dual-purpose pumper.  My feeling is that
these allowances for personal and property protection are
generous since they are in addition to the private security
and fire protection units provided by the refinery.

"It is possible that the refinery's atmospheric pollutants will require more frequent cleaning and painting of
public facilities such as park benches, playground equipment, etc., especially in the southern portion of the town.
We are presently spending $15,000 on these services.  In an
attempt to be safe, I have assumed that these outlays will
be doubled.

"This completes the background to Slide 3, the town hall
account under the hypothesis that the town does not reduce
present taxes with the refinery.  Given all the above assumptions, the present value at ten percent is about +$62
million.  The additional property taxes far outweigh the
additional cost of municipal services.

"If, on the other extreme, the town decides to maintain
municipal services at present levels, then we must refer to
Slide 4.  Under this option, the outlays for additional
street construction and maintenance, trash disposal, etc.
will be the same as in Slide 3.  When these are added to
the other municipal outlays required to maintain the present
level of services we obtain the 'tax to be raised' column in
Slide 4.  This figure, divided by the sum of the assessed
value of present town properties plus the refinery valuation
yields the projected tax rate.  This projected tax rate is
then applied to the refinery's valuation to obtain the
refinery property tax bill under the assumption that the

George Banks's Slide 4
Oronoco Refinery Impact on Tax Rate Assuming Same Level of Public Services Provided
With Refinery as Without (All Figures Except Tax Rate in Millions of 1974 Dollars)

| Year | Base Valuation | Oil Refinery Valuation | Total Valuation | Tax Raised (Includes Added Service Cost) | Tax Rate $/1000 | Oil Refinery Property Tax (Cash Flow) |
|---|---|---|---|---|---|---|
| 1974 | 47.728 | 0 | 47.7280 | 4.3770 | 100 | 0 |
| 1975 | 48.028 | 42.0000 | 90.0280 | 5.6526 | 63 | 1.323 |
| 1976 | 48.228 | 49.0000 | 97.2280 | 6.1924 | 64 | 3.136 |
| 1977 | 48.528 | 56.0000 | 106.5280 | 6.2661 | 59 | 3.304 |
| 1978 | 48.728 | 70.0000 | 118.7280 | 6.5676 | 55 | 3.850 |
| 1979 | 50.028 | 70.0000 | 120.0280 | 6.9129 | 58 | 4.060 |
| 1980 | 52.228 | 70.0000 | 122.2280 | 7.2284 | 59 | 4.130 |
| 1981 | 52.528 | 70.0000 | 122.5280 | 7.5377 | 62 | 4.340 |
| 1982 | 52.728 | 70.0000 | 122.5280 | 7.8922 | 64 | 4.480 |
| 1983 | 53.028 | 70.0000 | 123.0280 | 8.2635 | 67 | 4.690 |
| 1984 | 53.228 | 70.0000 | 123.2280 | 8.5649 | 70 | 4.900 |
| 1985 | 53.528 | 70.0000 | 123.5280 | 8.7833 | 71 | 4.970 |
| 1986 | 53.728 | 70.0000 | 123.7280 | 9.0847 | 73 | 5.110 |
| 1987 | 54.028 | 70.0000 | 124.0280 | 9.3861 | 76 | 5.320 |
| 1988 | 54.228 | 70.0000 | 124.5280 | 9.6875 | 78 | 5.460 |
| 1989 | 54.528 | 70.0000 | 124.5280 | 9.9889 | 80 | 5.600 |
| 1990 | 54.728 | 70.0000 | 124.7280 | 10.2889 | 82 | 5.740 |
| 1991 | 55.028 | 70.0000 | 125.0280 | 10.5904 | 85 | 5.950 |
| 1992 | 55.228 | 70.0000 | 125.2280 | 10.8919 | 87 | 6.090 |
| 1993 | 55.528 | 70.0000 | 125.5280 | 11.1934 | 89 | 6.230 |
| 1994 | 55.728 | 70.0000 | 125.7280 | 11.4959 | 91 | 6.370 |
| 1995 | 56.028 | 70.0000 | 126.0280 | 11.7974 | 94 | 6.580 |
| 1996 | 56.228 | 70.0000 | 126.2280 | 12.0999 | 96 | 6.720 |
| 1997 | 56.528 | 70.0000 | 127.5280 | 12.4016 | 98 | 6.860 |
| 1998 | 56.728 | 70.0000 | 127.7280 | 12.7033 | 99 | 6.930 |
| 1999 | 57.028 | 70.0000 | 128.0280 | 13.0050 | 102 | 7.140 |

level of municipal services is not increased with the refinery.  The annual tax bill is shown in the last column of
Slide 4.  Comparison of this column with the first column of
Slide 3 will reveal that the refinery will pay considerably
less property taxes if municipal services are maintained at
present levels.  The overall effect of opting for this alternative rather than maintaining tax rates at what they
would be without the refinery is to reduce the net present
value of the town hall account at ten percent from about $62
million to about $35 million.  Of course, it is not clear
that the town can usefully spend the additional $28 million
in present value on public services, but it certainly should
be aware of the decrease in Parable Beach income associated
with using the refinery's revenues to decrease the tax rate.
The town may very well decide to take some middle course,
reducing taxes less than implied in Slide 4, but increasing
public services less than envisioned in Slide 3, with intermediate results with respect to municipal income.

### The Labor Accounts

"The net set of accounts I examined involved the possible
changes in income of Parablites who would be employed in
either the construction or operation of the refinery.  Let's
start with the construction labor.  An estimate of the labor
skills required is shown in Slide 5.  A check with the local
unemployment office indicated that there are essentially no
Parablites in the higher skilled trades who are also unemployed, but that there are approximately 160 presently unemployed Parablites who would qualify as laborers and
helpers.  Discussion with the developers indicated that they
would like to hire Parablites but that I should check with
the relevant unions, which I did.  Union officials were adamant that:
1.  This was to be a union job.
2.  The unions had no intention of adding new members to
    their ranks.
A check of the Parablite unemployed revealed that not one of
them was a member of the subject unions.  My conclusion is
that at most a handful of presently unemployed Parablites

George Banks's Slide 5
Refinery Construction Skills Needed

| Skill | Number Needed |
|---|---|
| Pipefitter | 700 |
| Electrician | 250 |
| Insulator | 250 |
| Laborer | 200 |
| Ironworker | 150 |
| Boilermaker | 150 |
| Operating engineer | 100 |
| Millwright | 50 |

will be employed in the refinery construction. Those that
do will be runners and the like for the developer. The effect on town income of the construction payroll appears to
be nil.

"The case of the permanent operating personnel is somewhat different. For one thing, we don't have the problem
of already-established unions. For another, the developer
has promised to conduct a training program to prepare local
people for operating jobs. Finally, the developer has indicated that where possible, he will give preference to Parablites in hiring, and it's in his interest to do so.

"My estimate of the breakdown of the refinery labor force
is shown in Slide 6. Studies of similar projects elsewhere
indicates that the refinery will transfer in about fifty-five
top people from other installations. These people will occupy the bulk of the supervisory and key management jobs and
command the top salaries.

"With respect to the impact of the remaining 295 jobs on
Parable Beach income, it is of interest to study the structure of Parable Beach income. State unemployment figures
indicate that there are currently 400 Parablites unemployed
and seeking work. Forty-two percent of these people are
female, and thirty-five percent of these women have not completed high school. This leaves some sixty apparently
trainable women, which would be more than enough to occupy
the thirty or so clerical jobs to which they seem to be in
effect limited. In my calculations, I have assumed that all
these clerical jobs, paying an average of $7,000 after state
and federal income taxes, go to Parablites who would otherwise be earning nothing. This is a little extreme, but we
shall see that any errors introduced here are small overall.

George Banks's Slide 6
Refinery Labor Breakdown

| Category | Transfers | Locals* |
|---|---|---|
| General management | 10 | |
| Operations | | |
|   Supervisors | 30 | |
|   Operators | | 115 |
| Maintenance | | |
|   Supervisors | 5 | 10 |
|   Welders | | 15 |
|   Electricians | | 25 |
|   Mechanics | | 25 |
|   Pipefitters | | 25 |
|   Carpenters | | 15 |
| Engineering | 5 | 10 |
| Quality control | 3 | 10 |
| Fire and safety | | 15 |
| Accounting, personnel | 2 | 10 |

*Assumes six-month training program.

"The breakdown of the unemployed males, Slide 7, is interesting.  The state last-job categories are not very informative and I have not been able to get the State Unemployment
Agency to give me a simultaneous breakdown of age, education,
and skill, but roughly it appears that there might be something like 150 presently unemployed male Parablites who
could be trained for refinery operating and maintenance jobs.
I think it's a little too optimistic to think that, even
with preference, all of these people will land and keep refinery jobs.  In my estimates, I have assumed that 100 of
these jobs, averaging after training $9,000 per year after
state and federal taxes, will go to otherwise unemployed
Parablites.  These people would otherwise be on welfare.
The average value of welfare payments I have estimated at
$3,200 per year.  Thus, the increase in real income of these
100 people would be $5,800.  The remainder of the jobs go to
non-Parablites or Parablites who would be earning something
like $9,000 after taxes without the refinery.

"Under these assumptions, the annual increase in Parable
Beach employee income after start-up is $790,000.  Assuming
these people are paid at seventy-five percent of their working salary during a six-month training period commencing a
half-year prior to refinery operation, the present-valued
increase in Parable Beach employee income at ten percent is
$5,651,420 and at twenty percent is $2,587,340.  This estimate of the increase represents about twenty percent of the
refinery's gross payroll.

The Parable Beach Consumers' Account
"There has been considerable talk around town that the refinery will decrease the price or increase the quantity of oil
consumed by townspeople.  If this happened, it would

George Banks's Slide 7
Structure of Parable Beach Male Unemployment

| Skill (Last Job Held) | |
|---|---:|
| Professional | 8 |
| Sales and clerical | 5 |
| Craftsman | 45 |
| Operative | 67 |
| Non-farm labor | 115 |
| Age | |
| Under 22 | 36 |
| 22-35 | 59 |
| 35-45 | 61 |
| 45-55 | 37 |
| 55-65 | 35 |
| Over 65 | 12 |
| Education | |
| Less than 8 years | 32 |
| Less than 12 years | 85 |
| High school graduate | 74 |
| Less than 16 years | 41 |
| College graduate | 8 |

represent an increase in real income to Parablites. After
discussions with a number of industry officials, my judgment
is that any such effect is very likely to be insignificant.
The amount of oil consumed is rather insensitive to price,
so that even if there were a price change, there would be
little effect on quantity consumed. Further, in the absence
of price controls, in order to affect price, the refinery
would have to affect the cost of the most expensive oil consumed in the whole region. A 250,000 barrel per day refinery represents only a small proportion of the overall
regional consumption. With the refinery, the region would
still be consuming oil products from distant foreign refineries; this more expensive oil will determine the market
price. To put it another way, this refinery will have no
trouble selling all its output at the present market price
and, absent of controls, the refinery owner would be mad to
sell the oil for less. I suggest we not count on any gasoline or heating oil price decreases as a result of the refinery. There may be some decrease in retail price as a result of the shortening of truck hauls of present gasoline
and heating oil distributors who are now operating from ter-
minals in Metacomet, ten miles away, but this will not amount
to more than a cent or two a gallon. This would mean a savings of less than $50,000 per year. That is quite small
compared to some of the other numbers with which we are
dealing.[2]
------------------------------------------------------------
[2]More precisely, whether or not there will be downward pressures on local products prices will depend on whether or not
this refinery pushes all the products from the most expensive refinery source off the regional market. Assume for
the moment that Parable Beach is in New England. Then the
possible sources of refined products can be grouped into
five categories:
1.  Europe
2.  Eastern Canada
3.  U.S. Gulf
4.  Puerto Rico-Virgin Islands
5.  Mid-Atlantic
Assuming each of these areas had the capability of receiving
deep-draft tankers, a rough estimate of the differentials in
refiners' cost of landing products in New England relative
to those of a local refinery is given by:

| | Crude Transport | Tariff | Refining | Products Distrib. | Total |
|---|---|---|---|---|---|
| Europe | 0 | 55¢ | -20¢ | 40¢ | 75¢ |
| Eastern Canada/ Bahamas | 0 | 55¢ | -30¢ | 15¢ | 40¢ |
| Gulf | 5¢ | 0 | -15¢ | 45¢ | 35¢ |
| Puerto Rico/ Virgin Islands | -10¢ | 0 | -15¢ | 45¢ | 20¢ |
| Mid-Atlantic | 0 | 0 | 0 | 15¢ | 14¢ |

These numbers imply that if New England must import some of
its products from Europe the regional price of all products

will be determined by the landed cost of European fuel and,
absent of price control, the local refinery will be able to
make something like 75¢ per barrel in pre-tax profits. If,
on the other hand, domestic refining expands to the point
where all European products are pushed off the East Coast
market, then there will be downward pressures on prices
which could lead eventually to price drops of about 35¢ per
barrel. The following figure sketches the process.

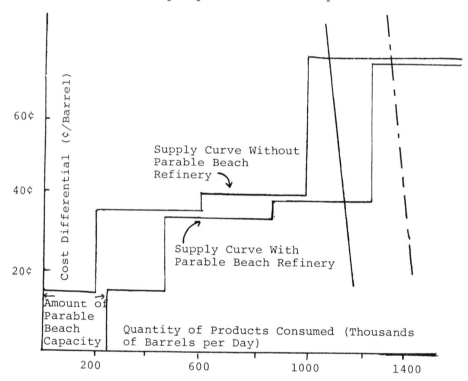

The addition of 250,000 barrels per day of local refining
shifts the supply curve to the right by this amount. If the
intersection of demand and supply was originally as that
shown by demand curve $D_1$ then this shift will force all the
most expensive products off the market and there will be
downward pressure on all regional products prices--not just
those produced by the refinery. If the East Coast was originally importing more than 250,000 barrels per day from the
most expensive source, as depicted by $D_2$, then the new refinery will not push all the highest-step oil off the market
and there will be no changes in prices even under free competition. Thus, the actual shape of the regional supply curve
of products becomes a very important factor in determining
whether there will be any price change. George Banks is assuming the situation is similar to that shown for the $D_2$
curve.

## The Parable Beach Landowners' Account

"The landowners' account raises an interesting problem.  The
local landowner most directly affected by the refinery is
Ephraim Ahab, who still maintains a legal residence in the
town.  Now we don't know what deal Mr. Ahab has made with
Oronoco, but knowing old Eph, we can be sure he is making
out all right.  My guess is in the neighborhood of $10,000
per acre for land that had a market value of perhaps $3,000
per acre until the refinery came along.  The difference is
about $4.6 million.  The question is, should we count this
increase in Ahab's income as an increase in Parable Beach
income?

"Under the rules with which we have been operating, we
probably should.  Ahab has been and is a legal resident of
the town and a long-time property owner.  Strictly speaking,
he qualifies.  However, for the last twenty years we have
only seen Ahab for a day or two a year, and it is quite
clear that his only interest in the town is this piece of
property.  He is not even a part-time resident, in the every-
day meaning, of the town.  I suggest we make an arbitrary
exception to our rules and not count him as a Parablite for
the purposes of municipal income, but I will be guided by
your judgment."

Banks looks questioningly at the Council and receives
seven affirmative nods or grunts in return.

"Okay, Ahab's increase in income is out.  Now Ahab is
not the only property owner in Parable Beach whose land
value might be affected by the refinery.  Fortunately, the
access road will all be on town-owned or refinery-owned land
so no private property takings are required.  However, property values in the neighborhood of the refinery and perhaps
in a wider area may be affected by the environmental and
aesthetic effects of the installation.

"The process units would be placed in the center of the
site so noise does not appear to be a problem.  However,
there will be some glare, some smell, and certainly some
visual impact.

"It is simply impossible to estimate with any degree of
accuracy how large the effect of these impacts on neighboring property values will be.  Oronoco claims these effects
will be insignificant and points to refineries elsewhere
which coexist more or less peacefully with residential
neighbors.  But I'm not so sure.  It's true that if people
turn out not to mind the refinery the property values won't
go down much.  But at this point it would be foolhardy to
predict how Parablites and the local property markets are
going to react to the refinery.  At the other extreme, it
could effectively preclude Parable Beach's being regarded as
a desirable residential community, just at the time when the
region appears to have reawakened to the town's unique location.

"Despite this uncertainty, one can obtain some insight
into the problem by examining Slide 8.  In this slide, I
have shown an estimate of the current market value of all

George Banks's Slide 8
Current Value of Neighboring Parable Beach Property

| Band (miles) | Acreage | Average Value/Acre | Total Value |
|---|---|---|---|
| Under 1/4 | 275 | $50,000 | $17.2 million |
| 1/4 - 1/2 | 360 | 48,000 | 21.8 million |
| 1/2 - 3/4 | 400 | 54,000 | 31.7 million |
| 3/4 - 1 | 100 | 60,000 | 7.5 million |
| 1 - 1-1/4 | 100 | 60,000 | 7.5 million |
| 1-1/4 - 1-1/2 | 120 | 54,000 | 6.5 million |
| 1-1/2 - 1-3/4 | 130 | 54,000 | 7.1 million |
| 1-3/4 - 2 | 150 | 48,000 | 7.2 million |

the town land within a quarter-mile of the refinery's boun-
daries, between a quarter-mile and a half-mile, and so on,
out to two miles. Many people feel that a refinery of this
size requires a site of a thousand or more acres in order to
allow for sufficient buffer zone. This would be about 400
acres more than the abandoned sand and gravel pit allows.
If one takes an extreme view and believes that as a result
of the refinery, the property within a quarter-mile of the
refinery will have zero market value, then the first element
of the last column represents an estimate of the loss in
Parable Beach income associated with this drastic drop in
prices. If one is even more pessimistic and believes that
all the land within a half-mile of the refinery will become
valueless, then the loss grows to $39 million.

"The point is that these numbers are not small. If realized, they would put a very healthy dent in the $62 to $34
million present value increase in town income we have com-
puted up to this point, depending on how severe and how
far-reaching the actual effect of the refinery on surrounding property values is.

"Further than this I cannot help you. I can only suggest
you visit some similar refineries elsewhere. And I might
suggest that if you do decide to accept the refinery you
either

1. obtain a clause in the agreement with the refiner
that he agree to buy any surrounding property at the present
price in real terms that is offered him through, say, the
next five years. If Oronoco really believes the effect will
be small, they will have little to lose by signing. Or,

2. set up a transfer system by which neighboring property owners whose property values do turn out to be nega-
tively affected can receive compensation for at least a
portion of the loss. This compensation might take the form
of, for example, complete abatement of property taxes or an
outright grant. As long as the town as a whole came out
ahead, as long as we actually experience an increase in real
municipal income, it is at least theoretically possible to
redistribute this increase so that everybody in the town
comes out ahead in terms of each Parablite's real income.

From the point of view of municipal income, 1 is preferable to 2, for in the first case the refinery will bear the loss in income associated with the drop in property values, while the second simply spreads the loss more evenly over the town.

"The problem is further compounded by the offshore crude terminal and its pipeline. The developer has already quietly made deals for pipeline right-of-way with the twelve private property owners along the route. While we do not know the amounts involved, since there was no compulsion, it's safe to assume that these property owners will be at least as well off with the line as without. I have conservatively assumed that they broke even. The developer is asking for a permit to build a portion of the line under town roads with the stipulation that the streets will be left in their original condition.

"The real problem is with the possibility of spills at the terminal, particularly a large spill, which later oiled the shore. I understand that Dr. Wayne will be giving you some input on the probability of such spills next meeting. For now, I simply want to mention that the possibility of spills will have some effect on shoreline property values. How much, I don't know. If a bad spill actually occurs, property values will be further depressed, again how much and for how long I don't know. Realtors in Santa Barbara interviewed after the large spill there estimated that shoreline property values dropped twenty-five percent in the areas directly affected after the spill and offered the opinion that in five years, assuming no more spills, these values would return to what they would have been without the spill. At a ten percent interest rate this would be equivalent to a present value loss of about six percent of the property's value. This figure was obtained by assuming that the property owner rents his property out at rents which are reduced proportionally to the transitory loss in property value. Under these assumptions, such a drop in property value would result in an equivalent one-short loss in town income of about $1.2 million at the time of the spill.

"Other losses in Parablite income in the case of a large spill coming ashore might result from actual cleanup expense. It should be borne in mind, however, that the great bulk of the cleanup expense will be borne by the oil company and the federal government and not the town. In this regard, you might want to read Mead and Sorenson's study of the economic impact of the Santa Barbara spill.[3]

"Other losses could accrue from the destruction of lobsters. There are presently eight full-time lobstermen and about six part-time lobstermen in the town plus three Parablites employed at the Bay Lobster Company. I estimate the take-home income of these Parablites at $150,000 per year.

---

[3]W. Mead and P. Sorenson, "The economic cost of the Santa Barbara oil spill," Proceedings Santa Barbara Oil Symposium, December 1970.

In the very extreme case that the lobstering was wiped out,
and these people could not find alternate employment, and no
portion of this loss could be recovered in the courts, an
upper bound of the present valued loss at ten percent interest rate would be about $1.4 million.  In reality, it is
unlikely that lobstering would be completely wiped out in
the case of even a very large spill.  Further, at least a
portion of any such loss should be recoverable from the oil
company.  However, once again the team might want to set up
some form of compensation system for those townspeople most
directly affected by a large spill.  Perhaps a portion of the
refinery property tax should be invested in a reserve fund
for this purpose.

Summary
"The last two slides are summaries showing my estimates of
the change in Parable Beach income associated with the refinery under a number of hypotheses about property tax policy
and neighboring land value effect.  These estimates are before respending.
    "With respect to respending, whatever the net municipal
multiplier is, it applies to the entire differential since,
*assuming the dump is the only other possible use of the
site*, these are true differentials.  However, since under
our assumptions the refinery puts a healthy dent in the unemployment in the town and since the numbers are so large
that there is no way any but a small portion of the differences can be respent within the town, we can be sure the
net municipal multiplier is quite small, especially if the
refinery doesn't have any significant depressing effects on
surrounding property values.
    "Remember also our assumption that the only alternative
use of the site is a town dump.  It is, of course, possible
that five years from now someone will come along who would
like to use the site for, say, a light manufacturing plant.
Having accepted the refinery, we will have effectively pre-
cluded ourselves from accepting this plant.  While the
present market value of the site indicates that people believe the likelihood of this happening is low, the possibility should enter your thinking.  I could work up some
other example alternatives, but frankly I ran out of time.
Thank you."
    Montfort:  "Thank you, George.  Any questions?"
    Beatrice Cenci speaks up.  She is a local lawyer and
insurance agent, thought to be leaning toward the refinery.
"Yes, I have one.  George, it seems to me you violated one
of your own rules on this property value business.  Back
in the Paragon Park analysis you pointed out that a drop in
price was not necessarily a loss in municipal income since
the loss in seller income would be matched by a gain in the
buyer's real income and as long as both buyers and sellers
were Parablites, this was a wash:  is that right?"
    George, warily:  "Yes."

George Banks's Slide 9
Summary of Accounts for Refinery Impact on Parable Beach (Millions of 1974 Dollars, 10%
Interest Rate)

| Account | Net Present Value | |
| --- | --- | --- |
| | No Decrease in Property Tax Rates | No Increase in Municipal Services |
| Δ Town Hall | +62.0 | +34.0 |
| Δ Parable Beach construction labor | nil | nil |
| Δ Parable Beach permanent employees | +5.6 | +5.6 |
| Δ Parable Beach oil consumers | +0.5 | +0.5 |
| Δ Parable Beach landowners | | |
| No adverse effects on neighboring property | 0 | 0 |
| Property values within 1/4 mile reduced to zero | -17.2 | -17.2 |
| Property values within 1/2 mile reduced to zero | -39.0 | -39.0 |

George Banks's Slide 10
Present Value of Increases in Parable Beach Income
Associated with Refinery Rather than Town Dump
(Ten Percent Interest, Before Respending, Millions
of 1974 Dollars)

|  | No Decrease in Taxes | No Increase in Municipal Services |
|---|---|---|
| No adverse effects on neighboring property | $68.1 | $40.1 |
| Property values within 1/4 mile reduced to zero | 50.9 | 22.9 |
| Property values within 1/2 mile reduced to zero | 29.1 | 1.1 |

"Well, then, what if the neighboring property values drop
and these properties are then picked up for a song by other
Parablites?  Then this change in property values would have
no net effect on municipal income, right?"

"Oh, I see what you're getting at," answers George, stal-
ling for time to collect his thoughts, "and I should have
spent more time on it in my presentation.  No, there is a
difference.  The key question is, is the object whose price
has changed still the same object as far as the Parablite
buyer is concerned?  For example, let's assume that as a
result of a shift away from beer drinking in the region, a
Parablite barowner finds he has to reduce his beer prices,
with a loss in income to himself of five cents a bottle.  A
Parablite customer who enjoys this beer in this tavern just
as much after the price drop as before gains five cents.
Then we have a wash.  On the other hand, let's assume that
there is a report that the brewery's beer is contaminated,
and the barowner is forced to lower his price a nickel to
get any business.  Now our Parablite customer is not sure
the report is true but as a result he is willing to pay less
for this beer.  If before, he was willing to pay forty cents
for the beer, he is now willing to pay only thirty-five
cents.  Then at forty cents he was just breaking even buying
the beer before the report, and at thirty-five cents he is
just breaking even after the report.  He has seen no increase in his real wealth.  The bartender's loss would not
be compensated, and we'd have a net drop in real municipal
wealth.

"In the case at hand, I have assumed that if the refinery
acts to lower the price of surrounding property it also
lowers the price that Parablites would be willing to pay for
this property.  To the extent that the drop in market value
is not matched by a drop in willingness to pay, then some
deductions from the property losses in the last slide are in
order.  Basically, it's a common-sense proposition.  Is the
object really the same object after the price change as

before?  Is a piece of land the same piece of land with a
refinery nearby as without?  This is the kind of question
that must be answered whenever one is dealing with a buyer-
seller transaction within the black box after a price
change."

Mrs. O'Houlihan:  "George, I have two questions.  One is
I thought you said that these real income analyses generally
indicate that the changes in black-box income are much lower
than usually claimed by proponents.  But here this doesn't
seem to be the case.  You're estimating that, if the refinery doesn't affect property values adversely and we don't
reduce taxes, the town could make over $62 million present
valued.  That's equivalent to handing every person in the
town $6,200 on a one-shot basis."

George:  "Yes, this is an exception.  And in my opinion,
the reason it's an exception is the ecological and aesthetic
pollution associated with the refinery.  Notice the $65 mil-
lion is almost all property taxes.  Refineries in our situation pay a lot more in property taxes than they cost the
town in monetary terms in additional municipal services,
while other types of development proposals usually break
even.  This is a reflection of the environmental disbenefits associated with a refinery.  If there were none, you
could be sure that towns all up and down the coast would be
falling all over each other trying to get a refinery.  Refineries would go from town to town asking who would give them
the lowest property taxes.  Tax abatement offers would begin
coming in in a hurry.  It wouldn't be long before the competing towns would knock the refinery's tax bill down to the
cost of additional public services.  Then, even from a
purely monetary point of view, the refinery would be a
break-even development as far as the town is concerned.
It's the environmental disbenefits which are restraining the
other towns from competing for the refinery by cutting their
property taxes.  I'm afraid the operation of the market
tends to almost always leave us with difficult choices."

"Thank you, George.  There's no such thing as a free
lunch, is there?  My last question is a simple one.  The
things you are saying seem to me to make sense.  Why in all
the economic analyses I have heard about or seen discussed
before this council have I never seen such things as real
black-box income?"

George:  "Well, partially it's lack of training.  Problems such as ours tend to be ignored by economists as too
trivial to be of interest.  And in truth, there isn't any
great secret to it.  Anybody who is reasonably careful to
keep track of both the plusses and the minusses can do it.
Partially, it's laziness.  Sometimes people who should know
better are simply too lazy to carry their estimates through
to the bottom line, the net effect on black-box income.  So
they'll make some estimates of number of jobs, gross payrolls and the like and stop there.  This is akin to going
before a corporate board of directors with an analysis of an
investment which indicates only what the investment's

revenues will be.  The immediate question would be, 'Yeah,
but how much money are we going to make?'  Unless you ad-
dressed the bottom line, you would be kicked out on your
tail.  Without similar tailkicking in coastal zone analysis,
people tend to get lazy and don't work things down to the
bottom line.

"Of course, the really important reason for the poor
quality of coastal zone economics is that invariably the
people who are doing the work are not doing it in order to
shed light on what should be done.  Rather, they have
already decided what should be done and they are doing the
work to convince someone else that what they want done
should be done.  This is not economics.  It is not analysis.
It is salesmanship."

Simon Montfort:  "Thank you again, George.  I don't think
you've made our decision any easier, but I'm sure you've made
it a better informed one.  The council will meet at eight
p.m. on Wednesday to receive the presentation on the environmental impacts of the refinery.  May I have a motion to
adjourn?"

## Chapter 6

## THE ORONOCO REFINERY FROM THE POINT OF VIEW OF THE STATE

### The Governor's Problem

The next morning, George Banks receives a call from his superior, the State Commissioner of Natural Resources, E. Kyle Renick.

"George, I understand you've been doing some work on the Oronoco refinery on the side."

"Yes, sir."

"Is the town really seriously considering awarding the necessary permits?"

"Yes, sir. I think they are."

"Hmm. I hadn't realized it had gotten that far. The governor just called me and expressed concern on the project and asked my advice. She's evidently beginning to get a lot of pressure from Horseham and some of the other towns in the area. Frankly, she doesn't know what to do and needs guidance. I understand you have just prepared a report for the Parable Beach town council and I thought we could give her your report."

"We could, but I don't think it would be appropriate. You see, my report addresses the real income dimension of the refinery from the point of view of the town of Parable Beach. The governor presumably is interested in the effect of the refinery on the real income of the entire state, which is quite a different matter. If she were to interpret the results for Parable Beach as applying to the state as a whole, we would be grossly misleading her."

"I'm not sure I see the difference. But the key question is how long it would take you to generate a report which would not be misleading."

"About a week."

"Okay, drop what you've been doing and get me this report. It will delay the Phase VI-B review of the comprehensive coastal zone land use plan a week, but since that plan has already been under review for two years, I don't suppose that will be critical."

"Yes, sir."

### George's Analysis for the State

Banks realizes he is now dealing with a new enlarged black box encompassing every citizen of the state. Once again he decides to accept as a postulate that the site can only be used as a town dump or for the refinery. Further, he decides to assume that if the refinery is not built in Parable Beach, it will not be built anywhere in the state. The basic point is that in doing black-box income analyses one must *explicitly lay out all the alternatives open to the black box*. Given these assumptions, George decides to use

Table 6.1
Accounts for Analysis of Refinery From the State's
Point of View

---

Δ State construction and operating employees
Δ Parable Beach Town Hall
Δ Horseham town hall
Δ State property owners
Δ State refinery owners
Δ State House
Δ State consumers
Δ Responding

---

a projection of the status quo as his baseline.  Relative
to this baseline, he sets up the accounts shown in Table
6.1.

George decides it's reasonable to assume that ninety percent of the construction employees will be state residents.
He already knows essentially all of them will be people who
are already members of the construction trade unions.  Talking with people at the state labor office, he finds that,
except for some seasonal fluctuations, there has been little
long-term unemployment among these union members.  The labor
officials offer the opinion that the refinery would put a
real pinch on the local construction labor market, causing
some other projects to be deferred or extended.  The officials also indicate that they think it quite unlikely that
the unions would open their ranks to presently unemployed
people in order to accommodate the additional demand.  On
the basis of these discussions, Banks decides that there
would be little difference in the take-home pay of the labor
actually utilized and thus the overall effect of the construction payroll on state income would be nil.

With respect to the operating personnel, George assumes
that all the 295 non-expatriate jobs would go to state residents and that on the basis of Oronoco's promise to train
and hire presently unemployed state residents, he can count
on 200 of these jobs with an average pay of $9,400 after
federal income tax going to otherwise unemployed people presently averaging $2,600 on federal welfare.[1]  Under these
assumptions, the annual increase in state income due to the
payroll will be equivalent to a present value increase of
$124 million at ten percent interest rate over twenty-five

---

[1]Notice that for the enlarged box, we now include the formerly unemployed employee's state income tax as part of his
income since when he transfers this income to the state it
stays within the black box.  We are also assuming that the
cost of state services does not rise as a result of this income tax payment.  By similar reasoning, we no longer count
the state welfare he was receiving as part of his pre-
refinery income since this was a transfer from other entities within the black box.

years. *This is a net increase if and only if the capital which would be invested in the refinery would not be invested in the state otherwise.*

The Parable Beach Town Hall account remains unchanged. George decides it is unlikely that Parable Beach will not reduce its tax rate markedly if the refinery is built and therefore uses the lower $34 million estimate for the present value of this account.

Horseham will be affected by the refinery in a number of ways. There will be additional travel, much of it truck traffic, especially on the road from the Parable Beach town line just south of the refinery to the interstate five miles away--all of this travel on Horseham town roads. Applying the data gathered in the Parable Beach study, Banks estimates the additional cost of this maintenance and repair to Horseham at $512,000 annually. Since the present terminals are located both closer to the center of the market and closer to the main highways, this is a net effect as far as the state is concerned rather than simply a shift in maintenance costs from one locality within the state to another.

The property owner account raises the same question as in the Parable Beach case relative to Ephraim Ahab. George decides that while Ahab is for all legal purposes a resident of the state, he will not count any increase in Ahab's income with the sale of the property to the refinery on the grounds that he is no more an actual resident of the state than he is of Parable Beach. With respect to the property owners who might be adversely affected, the analysis is the same as that for Parable Beach, except now Banks must include property owners on the Horseham side of the refinery as well. Doing so yields the losses shown in Table 6.2 as a function of the severity of the effect.

## The Refinery Owners' Account
Banks has taken the precaution of including a refinery owners' account in his list, since this refinery will enjoy transport cost and tariff advantages over the more distant refineries with which it will be competing which will translate into profits.[2] Insofar as state residents share in

Table 6.2
Losses in State Income as a Result of Decrease in Property Values (1974 Dollars)

| | |
|---|---:|
| No effect | 0 |
| Property values within 1/4 mile reduced to zero | -$54,000,000 |
| Property values within 1/2 mile reduced to zero | -$99,000,000 |

[2]Once again, "profits" in this context means returns to shareholders *above* what they could have received if they invested their money elsewhere, returns above the normal interest rate.

these profits, this will be an increase in income for the
people.  Therefore, Banks will have to obtain an estimate on
the refinery's profits.  He also needs this estimate to compute the state's take in corporate income tax.  Finally, the
refinery's profits are intimately linked with the price of
the refinery's products; this price, in turn, will be the
key input to the consumers' account.
    Through a search of the literature and discussions with
the developer and several independent refinery engineers,
Banks comes up with the estimates of the operation's outlays
shown in Table 6.3, exclusive of crude cost and state and
federal taxes.  The equivalent unit cost per barrel of output can be offered by solving for C in the following equation.

$$927,000,000 = \sum_{n=3}^{25} \frac{1}{(1 + i)^n} (C \cdot 233,000 \cdot 365 \cdot .95)$$

In this equation, 927,000,000 is the present value of the
outlay at ten percent interest rate, 233,000·365·.95 is the
yearly output in barrels, and i is the interest rate.  C is
the price per barrel the refinery would have to obtain to
just break even on its expenses exclusive of crude cost and
state and federal income taxes.  At ten percent this figure
is $1.48.  The yearly output of this refinery is about
eighty-five million barrels.
    On the revenue side, George realizes that, assuming no
price controls, the market price of the refinery's output
will be determined by the cost of the most expensive oil
products landed and consumed in the region.  With respect to
this *marginal* oil, there are several possibilities.  At present, the most expensive products consumed in the region are
those refined at European refineries.  This oil suffers both
from a 40¢ per barrel disadvantage in transport costs and
about a 55¢ per barrel disadvantage in tariff due to higher
tariffs on products than on crude.  The actual cost of refining is slightly less in Europe, about $1.25, so the European
refinery has an edge there.
    The next most expensive oil landed in the region is that
refined in closer foreign refineries in eastern Canada and
the Bahamas.  These refineries suffer the same tariff differential, about a 15¢ product transport differential, and
are about 30¢ per barrel cheaper in unit refinery cost.  The
next most expensive competitor is domestic Gulf Coast refineries, which suffer no tariff differential and are about
15¢ per barrel cheaper in actual refining but suffer a 45¢
per barrel product transport differential.  The next best
alternative is domestic Caribbean refineries which have an
edge over the Gulf.  The final possibility is a refinery
complex 500 miles down the coast, which suffers a 15¢ transport differential.  Differences in Persian Gulf crude cost
landed at the various refineries are not really significant,
generally less than 5¢ per barrel, assuming all the

Table 6.3
Refinery Account (Millions of 1974 Dollars)

| Year | Construction Costs | Operating Labor Costs | Vendor Costs* | Property Taxes** | Total Flow |
|------|------|------|------|------|------|
| 1974 | -25  | 0    | 0    | 0    | -25.0 |
| 1975 | -300 | 0    | 0    | -3.0 | -303.0 |
| 1976 | -75  | -30  | -20  | -3.1 | -128.1 |
| 1977 | 0    | -40  | -40  | -3.2 | -83.2 |
| 1978 | 0    | -40  | -40  | -3.8 | -83.8 |
| 1979 | 0    | -40  | -40  | -4.0 | -84.0 |
| 1980 | 0    | -40  | -40  | -4.1 | -84.1 |
| 1981 | 0    | -40  | -40  | -4.3 | -84.3 |
| 1982 | 0    | -40  | -40  | -4.5 | -84.5 |
| 1983 | 0    | -40  | -40  | -4.7 | -84.7 |
| 1984 | 0    | -40  | -40  | -4.8 | -84.8 |
| 1985 | 0    | -40  | -40  | -5.0 | -85.0 |
| 1986 | 0    | -40  | -40  | -5.1 | -85.1 |
| 1987 | 0    | -40  | -40  | -5.2 | -85.2 |
| 1988 | 0    | -40  | -40  | -5.5 | -85.5 |
| 1989 | 0    | -40  | -40  | -5.6 | -85.6 |
| 1990 | 0    | -40  | -40  | -5.7 | -85.7 |
| 1991 | 0    | -40  | -40  | -5.9 | -85.9 |
| 1992 | 0    | -40  | -40  | -6.1 | -86.1 |
| 1993 | 0    | -40  | -40  | -6.2 | -86.2 |
| 1994 | 0    | -40  | -40  | -6.4 | -86.4 |
| 1995 | 0    | -40  | -40  | -6.5 | -86.5 |
| 1996 | 0    | -40  | -40  | -6.7 | -86.7 |
| 1997 | 0    | -40  | -40  | -6.9 | -86.9 |
| 1998 | 0    | -40  | -40  | -6.9 | -86.9 |
| 1999 | 0    | -40  | -40  | -7.1 | -87.1 |

Note:  Negative numbers again represent outlays.

*Purchased goods and services.
**Assuming tax rate is decreased.

contenders are served by a deepwater terminal.  Table 6.4
summarizes the situation relative to the competitors.

The "Total" column represents the unit profits above
those required to obtain a normal return on invested capital
available to the refinery depending on who its marginal competitor is.  For example, if the region is forced to import
some of its products from European refineries, this differential is estimated at 75¢ per barrel or about $60 million
per year in profits in excess of normal return on capital.
However, if refining capacity in eastern Canada or the Bahamas expands to the point where all European products are
forced off the regional market, assuming effective competition, the price of products could drop about 35¢ per barrel
and the excess profits would drop to about 35¢ per barrel,
or about $30 million per year.  In this context, the term
"excess profits" is used not in a pejorative sense but
rather in a technical sense--profits in excess of the interest rate the refinery's investors could obtain elsewhere.
Thus, excess profits relate to a true increase in investor
income relative to what the investor could have had without
this project.  If domestic refining capacity expands to the
point where all imported products are forced off the regional
market, then the differential drops to perhaps 25¢ or even
15¢ in the extremely unlikely case that any excess of refining capacity in the Middle Atlantic region occurs.

Inquiries to Oronoco indicate that state residents own
two percent of the company, and thus two percent of the
after-corporate-income-tax profits will go to people within
the black box.  Also, the state corporate income tax is six
percent, which yields the estimates of the increase in state
income associated with shareholder profits and state corporate tax revenues shown in Table 6.5 as a function of marginal competition.

George assumes that additional State House expenses associated with the refinery will match the additional expenses
suffered by the Parable Beach and Horseham Town Halls, which
come to $9.1 million present valued at ten percent for
twenty-five years.  George is not too worried about errors
introduced by this rather casual assumption, since he can
already see that any error introduced will be small in the
overall picture.

## The State Oil Consumers' Account

The state oil consumers' account is, in a sense, the inverse
of the refinery owners' account.  If the price does not
drop, the refinery owners' take is maximized and the consumers see no increase in real income.  To the extent that the
price of products does drop, the real income of the refiners
decreases and the income of the consumers increases.  There
are, however, two complicating factors:
1.  Not all refinery owners are state residents.
2.  A products price decrease, if it occurs, will decrease
the price of all products consumed in the state, not just
those produced by this refinery.

Table 6.4
Estimates of New England Refinery Vs. Outside Refinery, Developer Differentials
(Price Per Barrel)

| Marginal Competitor | Δ Crude Transport* | Δ Tariff** | Δ Refinery | Products Distribution | Total |
| --- | --- | --- | --- | --- | --- |
| Europe | 0 | 55¢ | -25¢ | 40¢ | 75¢ |
| Canada/Bahamas | 0 | 55¢ | -30¢ | 15¢ | 40¢ |
| Gulf of Mexico | 5¢ | 0 | -25¢ | 45¢ | 35¢ |
| Puerto Rice/Virgin Islands | -10¢ | 0 | -15¢ | 45¢ | 20¢ |
| Mid-Atlantic | 0 | 0 | 0 | 15¢ | 15¢ |

*Assumes deep-draft terminal in all six refinery locations, including Parable Beach. Marginal crude is Persian Gulf oil.
**41¢ per barrel crude-products differential plus 20¢ per barrel crude tariff.

Table 6.5
Changes in State Income Associated With Shareholder Profits and State Corporate Income Tax, and Additional State House Expenses (Millions of 1974 Dollars, 10% Interest Rate)

| Marginal Competitor | Estimate of Present Value Pre-Tax Profits | Present Value State Income Tax | Present Value of State Shareholder Profits | Δ State House Expenses |
|---|---|---|---|---|
| European refinery | 555.5 | +33.6 | +10.9 | -9.1 |
| Bahamas/Canada | 303.6 | +18.1 | +6.4 | -9.1 |
| Gulf Coast | 192.4 | +11.8 | +3.6 | -9.1 |
| Mid-Atlantic | 116.2 | +7.3 | +1.8 | -9.1 |

The first problem we have already handled in the refinery
owners' account. Tackling the second, George knows that
currently 600,000 barrels per day of oil are being consumed
in his state. He doesn't know how much people in the state
are going to consume in the future, but he notes that the
lowest projections put out by the State Energy Office call
for no growth through the next twenty-five years, while the
highest call for a four percent annual growth rate. He decides to study these two extremes as a function of who the
marginal competitor is. The results are shown in Table 6.6
for a ten percent interest rate. George notes that these
numbers can be quite large. He also knows that the entire
East Coast is presently importing somewhere in the vicinity
of 350,000 barrels per day of oil from Europe, so the refinery is unlikely to push all this oil out of regional markets, even at zero percent growth, and still less likely to
keep European oil off the market in the face of a four percent consumption growth. There is certainly no chance a
single refinery can push not only all the European oil off
the East Coast market but also all the Canadian-Bahamian
oil, and George decides to no longer consider the Gulf, domestic Caribbean or mid-Atlantic as potential marginal competitors. Finally, George knows that even if all the
European oil were forced off the market, the impediments to
competitive forces, especially in the gasoline markets,
would make the price drop a gradual and possibly nonexis-
tent phenomenon. Nonetheless, Table 6.6 makes the rather
obvious point that if a products price drop could somehow
be effected, the increase in state income associated with
this drop could be quite large and certainly much larger
than the resultant loss in state refinery owners' income and
state tax revenues with the price drop.[3]

Table 6.6
Present Valued Increase in State Oil Consumer Income as a
Function of Twenty-Five Year Price Drop and Future Consump-
tion Growth (Ten Percent Interest Rate, 1974 Dollars)

| Marginal Competitor | Drop in Prices (Dollars/ Barrel) | Present Value Increase in Consumer Income at 0% Growth | Present Value Increase in Consumer Income at 4% Growth |
|---|---|---|---|
| Europe | .00 | 0 | 0 |
| Bahamas/ Canada | .30 | +$482 million | +$844 million |
| Gulf | .45 | +$723 million | +$1267 million |
| Mid-Atlantic | .60 | +$964 million | +$1690 million |

[3]We are assuming there are no other refineries already in
the state which would also be affected by the price drop and
that state residents do not own extra-state refineries which

## Putting It All Together Again

Having culled all this data, George is in a position to prepare his summary, Table 6.7, to present to the governor his
estimates of the direct increase in state income associated
with opting for a refinery in Parable Beach as opposed to
the status quo as a function of both the refinery's marginal
competitor and the extent of the effect on neighboring property values.  The table shown gives only the results of
George's analyses for the low consumption growth hypothesis.

George notes the increase in consumer income with a price
drop, if it occurs, is the overwhelmingly large figure.
George also knows that the price drop is a very iffy proposition depending on both the event that this refinery pushes
all the European oil off the market and the event that there
is sufficient competition among refiners to drive the price
of all products down by the drop in the cost of the marginal
oil.  George realizes the individual petroleum products markets vary in the degree to which competitive pressures are
allowed to operate.  In the residual fuels market, there is
considerable competition.  The buyers, utilities and large
industrial plants, are individually strong, well-informed
shoppers who can obtain their needs from a large number of
alternative sources if local suppliers do not offer competitive prices.  Brand names mean nothing.  Historically,
prices have been quite flexible.  At the other extreme is
the gasoline market.  The majors are vertically integrated
down to the retail level, controlling prices directly in
eighty percent of the retail market.  Brand preference in
consumers can be and is being maintained by intensive adver-
tising.  Prices in the majors' outlets are relatively in-
flexible, set on a centrally administered basis, and
historically have not been responsive to momentary changes
in supply and demand.  In short, George chooses to regard
the consumer savings in Column 2 of Table 6.7 as an upper
bound, a best-of-all-possible-worlds estimate, and to concentrate on the no-price-drop (leftmost column) case.  Nonetheless, upper-bound numbers such as the consumers' savings
under general price drop are insightful in that they point
us toward potential areas of large increases in state income
and thus have important, obvious implications for state policy.

In the leftmost no-price-drop column, the total present-
valued numbers range from a high of +$86 million (no adverse
effect on neighboring property) to a low of -$13 million.
If these changes were spread evenly over the state, the upper number would be equivalent to handing each of five million citizens in the state about $47 on a one-shot basis;
the lower number is equivalent to taking away $2.50.  Of

---

would be adversely affected.  As can be seen from the relative magnitudes, even if all the state's products were refined internally, these effects would be much smaller than
the increase in consumer income.

Table 6.7

Summary of accounts for refinery impact on state before responding (Net present value, 25 Years, Millions of 1974 Dollars, 10% Interest Rate, Low Consumption Growth)

| Account | Marginal Competitor | |
|---|---|---|
| | Europe | Bahamas/Canada |
| Δ State construction and operating employees | +12.4 | +12.4 |
| Δ Parable Beach Town Hall | +34.0 | +34.0 |
| Δ Horseham Town Hall | -4.6 | -4.6 |
| Δ State property owners | 0 | 0 |
| Δ State refinery owners | +10.9 | +6.4 |
| Δ State House | +24.5 | +9.0 |
| Δ State consumers | 0 | +482.0* |
| (Note: The above assumes no adverse effect on neighboring property.) | | |
| Δ Total | +77.2 | +548.3 |
| Property values within 1/4 mile reduced to zero | -54.0 | -54.0 |
| Δ Total | +23.2 | +485.2 |
| Property values within 1/2 mile reduced to zero | -99.0 | -99.0 |
| Δ Total | -21.8 | +443.4 |

*This column assumes competition sufficient that prices drop to cost of marginal products. If this is not the case, then the numbers in the leftmost column hold even if no European oil is landed.

course, these changes will not be spread evenly over the
state. In fact, as long as no price drop occurs, a large
portion of the changes will be concentrated in Parable
Beach, as we've already seen. And even within Parable Beach
these changes will not be spread evenly unless an explicit
system for compensating abutting property owners is set up.

In general, as one moves to larger and larger black boxes,
the average unit effect of a given project on market wealth
becomes smaller and smaller, despite the fact that the aggregate effect is larger. This presents a very difficult
problem for our political system. It will pay a small group
of people, who on a per-unit basis are affected severely, to
join together and devote time and money to lobby for the
particular alternative favorable to their parochial inter-
ests. On the other hand, it will not pay a much larger,
more diffuse group to make the same individual effort, despite the fact that, in aggregate, their loss in market
wealth may be much greater than the gain in income to the
smaller black box. Hence the frequency of decisions at
every governmental level which are inconsistent with total
black-box income at that level.

The above figures refer to the direct increases in income. These increases in income will, of course, be respent
in a variety of markets, some of which will be within the
state. As we argued earlier, a generous estimate of the
average difference between the market price and what these
resources could earn in other employment would be twenty
percent of the resource's values. Such a number would imply
a much higher ratio of market price to alternate opportunity
value for labor, for non-labor resources are rarely subject
to unemployment.

An infinite net multiplier chain based on an average
twenty percent differential between market price and opportunity value would add twenty-five percent to the above final summary figures; a thirty percent differential would
tack on forty-three percent to the above figures; a ten percent differential would tack on eleven percent. Thus, if we
concentrate on the leftmost, no-price-decrease column of
Table 6.7, then as a function of the net multiplier assumed,
we have the final numbers shown in Table 6.8. Once again,
the policy implication of the increases in income associated
with respending, whatever the actual value of the net multiplier, are hardly striking.

Table 6.8
Impact of Varying Net Multiplier on State Income for the
No-Price-Decrease Hypothesis

| | Average Percentage Differential Between Market Price and Opportunity Value in Respending Markets | | | |
|---|---|---|---|---|
| | 0% | 10% | 20% | 30% |
| No change in neighboring property values | +77.2 | +85.7 | +96.5 | +110.2 |
| Property values within 1/4 mile reduced to zero | +23.2 | +25.8 | +29.0 | +33.1 |
| Property values within 1/2 mile reduced to zero | -21.8 | -24.2 | -27.3 | -31.1 |

## Chapter 7

## THE REFINERY FROM THE POINT OF VIEW OF THE NATION

It is one week after George has submitted his report to the governor. George is having his regular monthly meeting with the regional coordinator of the Office of Coastal Environment in Washington, Martin Abkowitz. They have broken for lunch and the topic of the Parable Beach refinery has come up.

George: "Well, I don't know what's going to happen. My guess is that Parable Beach may well approve it but that the neighboring towns, who will not share in the property tax income, will be able to muster enough strength to block it at the state level."

Martin: "Where does the national interest lie in this?"

George: "I don't know. All the work I've done on the issue has focused exclusively on either the real income of Parable Beach or the real income of the state."

Martin: "Can't you tell what the real income impact on the country will be from that?"

George: "Not really. I'd have to go back and rework the figures to get the national income impact."

Martin: "I'd appreciate it if you would. We've been having quite a discussion of the national interest clause in the Coastal Zone Management Act and the Secretary is thinking about taking a position on some of these large-scale projects."

George: "Well, national income is only one dimension of the national interest."

Martin: "I know, but I still think it would be useful."

George, wearily, and mentally cursing Mrs. O'Houlihan: "Okay." George's office is funded largely by federal coastal zone management funds. He is in no position to refuse a request such as this. "I'll have it to you in a week."

George goes back to his office and begins the process anew.

FIRST STEP: What is the black box?
Answer: All U.S. citizens.
SECOND STEP: What are the alternatives to be analyzed?
George decides to look at:
1. Additional domestic refining in Parable Beach.
2. No refinery in Parable Beach; instead, equivalent amount of products is imported from Bahamian refinery. Capital which would have been invested in Parable Beach refinery is invested in Bahamian refinery.[1] Site remains town dump.

------------------------------------------------------------

[1]As we shall see, it is extremely important that we specify exactly what happens in the absence of the project directly under analysis. In the case at hand, it is reasonable to assume that the amount of products consumption is independent

Table 7.1
Accounts for National Income Analysis

---

Δ U.S. construction and operating employees
Δ Parable Beach Town Hall
Δ Horseham Town Hall
Δ State House
Δ U.S. consumers
Δ U.S. property owners
Δ U.S. refinery owners
Δ Federal government
Δ Respending

---

THIRD STEP:  What is the baseline?
Answer:  Status quo, Alternative 2 in second step.
FOURTH STEP:  Set up accounts.
George decides to break down his accounts as shown in Table
7.1.

## The U.S. Construction and Operating Employees

The refinery construction and operating employees are the
same as in the case of the state income analysis, except now
all the employees including the people who will be transferred to Parable Beach by Oronoco count as long as they are
U.S. nationals.  However, if as before, the actual construction labor is already near full employment, the same thing
will be true of U.S. manufacturers of process equipment.
The fifty-five people who are being transferred are highly
skilled, fully employed people who, through promotions, will
see some increase in income as a result of the refinery.
However, this is unlikely to be more than one or two thousand dollars per person.  George estimates the total at
$50,000 per year.  Except for this, then, the increase
through employment is the same as in the state case.  However, George now counts the formerly unemployed operating
personnel's increases at their gross average pay, $10,000
per year with no deductions for federal income taxes or
federal welfare, for these are transfers within the enlarged
black box.  More precisely, the loss of welfare payments and
the loss to income taxes of these people would show up as
plusses in the federal government account--if we had worked
everything through carefully.

Notice that Banks's assumption that, if the refinery is
not built at Parable Beach, the released capital would be
invested outside the country, is critical to this account.
If Banks had assumed that the capital would be invested inside the country in some other use, then that use would undoubtedly have required some U.S. labor.  And if that labor
were unemployed, the increase in income of that labor would

---

of whether or not there is a refinery in Parable Beach, in
which case our non-Parable Beach alternatives must be consistent with this assumption.

have had to have been deducted from this account.  Since refining is a very labor-extensive business, it is quite possible the net effect of investing in refining rather than
some other domestic capital improvement on employee income
would be negative.

## The Town Hall and State House Accounts
The Town Hall accounts are unchanged from the state analysis.
Banks assumes that no other town halls are significantly involved.  Once again, the assumption that the capital would
not be invested domestically if the refinery is not built is
critical.  The State House account is also unchanged.  Thus,
we have from Table 6.5, the figures in Table 7.2.

## The U.S. Property Owners' Account
The property owners' account is the same as in the state
case, except now there is no ducking our friend Mr. Ahab.
On a long shot, George calls Ahab in Florida and is mildly
surprised to find that Ahab is quite willing to tell him the
deal he has made with the refinery.  It's $12,000 per acre.
Ahab opines that if the refinery did not buy the land, he
could still sell it for $5,000 per acre.  George adds the
differential to the property owners' account.

## The U.S. Consumers' Account
Whether or not there will be any impact on U.S. consumer income with the building of the Parable Beach refinery will
depend on whether or not there will be any products price
changes.  George decides to examine two cases:
    1.  The Parable Beach refinery is not large enough to
force all the most expensive European oil off the East Coast
market.  In this situation, there will be no change in consumer income.
    2.  The refinery forces all European oil off the market
and competitive forces are strong enough to drive the products prices down to the landed cost of the next most expensive alternative source, Canadian/Bahamian refineries--a
drop in landed cost which George estimates at 30¢ per barrel.
The price drop in Case 2 would affect at a minimum all East
Coast consumers.  However, George realizes that if all the

Table 7.2
Δ Town Hall and State House Accounts (Net Present Value,
Millions of 1974 Dollars, 10% Interest Rate)

| Marginal Competitor | Parable Beach Town Hall | Horseham Town Hall | State Income Tax | State House Expenses |
|---|---|---|---|---|
| Europe | +34 | −4.6 | +33.6 | −9.1 |
| Bahamas/ Canada | +34 | −4.6 | +18.1 | −9.1 |

other refineries which will serve the East Coast while a
Parable Beach refinery is in existence are owned by Americans, then the increase in income to consumers of products
not produced in Parable Beach will be matched by decreases
in income in gross excess American refiners' profits.  From
the point of view of the national black box, the overall effect will be a wash.[2]  Thus, George can concentrate entirely
on those Americans who would consume the output of the Parable Beach refinery.  From his earlier work on the state
problem, he already knows that the present-valued increase
in consumer income with a 30¢ per barrel drop in prices at
ten percent over twenty-five years is $246.9 million.  Thus,
for the consumers' account, he has the figures in Table 7.3.

The Refinery Owners' Account
George already knows that the Parable Beach refinery is
likely to be a rather profitable investment for its owners,
thanks largely to the tariff differential.  Thus, this refinery's owners could enjoy some substantial increases in
income.  With respect to Parable Beach refinery ownership,
Banks decides to study two extremes:
1.  An entirely U.S. citizen-owned Parable Beach refinery.
2.  An entirely foreign-owned Parable Beach refinery.
Oronoco is practically all American-owned, but Banks believes
it will be instructive for the Secretary of Commerce to see
the results for a foreign-owned refinery.
    Table 7.4 shows an estimate of the net present value of
pre-tax profits for the Parable Beach refinery.  Deducting
estimated state and federal corporate taxes, which will be
counted in other accounts, leaves the shareholders the
amount shown in the rightmost column before federal individual income taxes.  If the shareholders are all foreign
and reciprocity laws allow them to escape individual income
taxes, then the increase in U.S. shareholder income is obviously zero.  If, on the other hand, the shareholders are
all American, then from what they could have earned on the
Parable Beach refinery we must deduct any excess profits
which they would have earned on the Bahamian refinery which

Table 7.3
Changes in U.S. Consumer Income* (10% for 25 Years, Millions
of 1974 Dollars)

| | |
|---|---:|
| No change in price | 0 |
| 30¢ per barrel drop in price | 246.9 |

*Net of consumers of non-Parable Beach refined products, for
the reasons given above.

------------------------------------------------------------

[2] If some of these other American-owned refineries are located in foreign countries, foreign taxpayers may, depending
on local tax laws, bear a share of the decrease in the profits of these refineries.  To the extent that this happens,
the net impact will not be a complete wash.  George decides
to ignore this effect.

Table 7.4
Disposition of Refinery Profits (Millions of 1974 Dollars,
10% Interest Rate)

| Marginal Competitor | Estimate of Present Value Pre-Tax Profits | State Taxes | Federal Taxes | Shareholder Net Present Value |
|---|---|---|---|---|
| Europe | 555.5 | 33.6 | 250.3 | 271.4 |
| Bahamas/ Canada | 308.6 | 18.1 | 139.4 | 151.1 |

otherwise would have been the target of their capital.  If
they had not invested in the Parable Beach refinery, then
the Bahamian refinery would have been earning 30¢ per barrel
in excess profits or some $246.9 million present valued.  If
we assume the Bahamian tax structure is such that it takes
sixty percent of these profits in excise duties, then the
shareholders would retain about $10.0 million, which as
long as they don't repatriate it, is free of U.S. tax.
Thus, from the rightmost column in Table 7.4 we need to deduct $100 million to obtain Table 7.5 for the U.S.-owned
refinery alternative.
    Notice that once again our assumptions about what happens
if the project doesn't take place have to be as complete as
our assumptions about what happens if the project does take
place.  If we had assumed, for example, that the refinery
owners invested their capital in a European refinery, then
we would obtain rather different numbers for this account.

The Federal Government Account
The federal government account will be affected in four principal ways by the refinery as opposed to the status quo and
capital otherwise invested in a Bahamian refinery.
    1.  There will be an increase in federal private income
taxes on formerly unemployed employees' and domestic shareholders' earnings, and a decrease in welfare payments to
formerly unemployed employees.  All these effects have already been counted by basing the employees' and refinery
shareholders' earnings increases pre-federal individual income tax rather than after federal tax.  We don't want to
count them again.
    2.  There will be an increase in federal pollution moni-
toring expense.  Banks roughly estimates this at $130,000
per year.
    3.  From Table 7.4, the refinery owners will pay federal
corporate income taxes the present value of which is $250.5
million if the marginal competitor is a European refinery,
or $139.4 million if the Parable Beach refinery pushes all
European oil off the regional market and competition forces
prices down to the landed cost of Bahamian oil.
    4.  In a situation where the refinery is replacing for-
eign refined products, there will be a change in the federal

Table 7.5
Δ Refinery Owners* (Net Present Value, Millions of 1974
Dollars, 10% Interest Rate)

| Marginal Competitor | Non-U.S.-Owned | U.S.-Owned |
|---|---|---|
| Europe | 0 | +171.4 |
| Bahamas/Canada | 0 | + 51.4 |

government's petroleum tariff revenues.  We have seen that
under the assumption of European or near foreign marginal
competition, the refinery's profits are largely the result
of a 55¢ per barrel differential in crude oil tariffs versus
refined product tariffs.  In these situations, with the refinery, the country will be importing an additional 250,000
barrels of crude per day and importing 233,000 barrels less
of products.  Currently (1974) the federal tariff on products is 62¢ per barrel, while that on crude is 21¢ per barrel.  On top of this, present federal policy allows a new
domestic refinery a five-year tax holiday on crude tariffs.
Hence, without the refinery, federal tariff revenues on
products which would have been imported in the absence of
the refinery are given by

$$\sum_{n=3}^{25} \left(\frac{1}{1 + i}\right)^n .62(\$/bbl) \cdot 233,000(bbl/day) \cdot 365(days/year)$$

which at a ten percent interest rate equals $359 million.
With the refinery and given the five-year holiday, we have
crude tariff revenues of

$$\sum_{n=8}^{25} \left(\frac{1}{1 + i}\right)^n .21(\$/bbl) \cdot 250,000(bbl/day) \cdot 365(days/year)$$

or about $81 million.  Thus, the provision of this domestic
refinery would involve a loss of about $278 million in pre-
sent value federal tariff revenues.
    When the refinery is replacing foreign products, much of
the refinery's profits are not a true increase in national
income, but merely a shift from tariff revenues to private
profits and the taxes on these profits.  Of course, if the
refinery were displacing domestically refined products in
the local markets--the case where the marginal competition
is Gulf Coast or Mid-Atlantic facilities--then there would
be no change in tariff revenues.
    Adding the change in tariffs to the pollution control ex-
pense and federal corporate taxes leads to the figures in
Table 7.6 for the federal government account.  Remember, we
have counted the increase in federal shareholder earnings
income and personal income taxes in other accounts.

Table 7.6
Δ Federal Government (Net Present Value, Millions of 1974
Dollars, 10% Interest Rate)

| Marginal Competitor | Loss in Tariffs | Pollution Expense | Corporate Income Tax |
|---|---|---|---|
| Europe | -281 | -1 | +151.4 |
| Bahamas/ Canada | -281 | -1 | +31.4 |

Clearly, the federal government qua government will end
up a loser on net as a result of the Parable Beach refinery
due to the shift in tariff revenues to private profits and/
or decrease in price.

The Respending Account
If the alternate investment to the Parable Beach refinery
were a domestic refinery in an area that had about the same
unemployment levels as the Parable Beach region, then the
respending account would be a complete wash from the point
of view of the national black box.  The net increase in national income associated with respending would be nil.  However, since we have assumed that, if the capital is not
invested in Parable Beach it will be invested outside the
country and if we further make the assumption that an insignificant portion of the Bahamian refinery's costs will be
spent in domestic markets with substantial unemployment,
then the respending computation is quite similar to that in
the case of the state.  One must make an estimate of the
average difference between price and value of alternate opportunities in the domestic markets where respending occurs
and add that differential to <u>all</u> the increases in national
income.

As noted earlier, if one makes the generous assumption
that on the average all goods in which respending occurs are
twenty percent overpriced, then the effect of this net multiplier will be to add twenty-five percent to all the final
changes in income.  George decides that this effect is not
particularly striking in view of all the other uncertainties
he faces.  He decides to complete his analysis on a pre-
respending basis and note that fact in his report to the
Office of Coastal Environment.

Putting the Whole Thing Together a Third Time
Combining all the accounts leads to Table 7.7, and Table 7.8
summarizes Banks's estimate of the increase in national income depending on who is the marginal competitor and whether
or not the facility is domestically owned.  Banks notices
that the latter possibility is extremely important to overall national income, especially if products prices don't
drop.  In this case, most of the increase in *national* income
is made up of the refinery profits less any decrease in

Table 7.7
Summary of Accounts for Refinery Impact on Nation (Net Present Value, 25 Years, Millions of 1974 Dollars, 10% Interest Rate)

| Refinery | U.S. Construction and Operating Employees | Parable Beach Town Hall | Horseham Town Hall | State House* | U.S. Consumers** | U.S. Property Owners† | Federal Government* | U.S. Refinery Owners Including Shareholders' Income Tax |
|---|---|---|---|---|---|---|---|---|
| U.S.-owned | | | | | | | | |
| Europe | +18.6 | +34 | -4.6 | +24.5 | 0 | +4.6 | -130.64 | +171.4 |
| Bahamas/Canada | +18.6 | +34 | -4.6 | +9.0 | +246.9 | +4.6 | +51.4 | +51.4 |
| Foreign-owned | | | | | | | | |
| Europe | +18.6 | +34 | -4.6 | +24.5 | +246.9 | +4.6 | -130.64 | 0 |
| Bahamas/Canada | +18.6 | +34 | -4.6 | +9.0 | +246.9 | +4.6 | -250.64 | 0 |

*Federal individual income taxes included in refinery owners' account.

**This column assumes prices drop to landed cost of Bahamian/Canadian products.

†This column assumes no effect on neighboring property. If property values within one-quarter mile are reduced to zero, subtract 39.4. If property values within one-half mile are reduced to zero, subtract 99.4.

Table 7.8
Direct Increase in National Income Associated with Parable
Beach Refinery (Net Present Value, Millions of 1974 Dollars,
10% Interest Rate, 25 Years)

| Marginal Competitor | U.S.-Owned | Foreign-Owned |
|---|---|---|
| No loss in property values | | |
| Europe | +117.9 | -53.5 |
| Bahamas/Canada* | +109.3 | +57.9 |
| Property values within one-quarter mile reduced to zero | | |
| Europe | +78.5 | -92.9 |
| Bahamas/Canada* | +69.9 | +18.5 |
| Property values within one-half mile reduced to zero | | |
| Europe | +18.5 | -152.9 |
| Bahamas/Canada* | +9.1 | -41.5 |

*Assumes price drops full difference in landed cost of
European products versus Bahamian/Canadian products.

tariffs.  This is not true of Parable Beach income or state
income, since both these latter entities experience a minute
proportion of the profits (with the exception of state income taxes, which are not affected by nationality of ownership).

One result of this dependence is that in the case of a
foreign-owned refinery in Parable Beach replacing Bahamian/
Canadian refined products, national income is estimated to
be decreased by close to $53 million on a present value ba-
sis, before any loss in neighboring property values, while
in the same situation, both Parable Beach and state changes
in income are positive.  In such a case we have a clear conflict between national income and state and municipal income.  On the other hand, consider the case of a U.S.
investor-owned refinery under the very severe assumption
that all property values within one-half mile are destroyed,
and assume the price doesn't drop.  National income is increased, due primarily to investor profits and federal taxes
on these profits.  At the same time, state and local income
are decreased because the smaller black boxes experience
only a small proportion of the federal taxes on investor
profits, but all the losses due to the effect of the refinery on surrounding values.  This example makes two very important points:
1.  In any income analysis, you must specify whose income
you're talking about.
2.  Often the conflicts between developer, municipality,
state, and federal levels, which invariably is labeled "environmental", is explainable in income terms.  This raises
the interesting and rarely explored alternative of compensation as a means of resolving some of these conflicts.  Con-
ventional land use legislation calling for "balanced,
comprehensive planning" will not make these conflicts disappear.
But that's another story.  Right now, George Banks is simply
happy to breathe a sigh of relief as he puts the finishing
touches on Table 7.8, for he realizes that, since the United
Nations is certainly too distracted to worry about this investment's impact on world wealth, he is finally through.

## CHAPTER 8

## A CHECKLIST

Our visit to Parable Beach is over.  Perhaps the single most important thing that we can take away from this sometimes difficult and tedious trip is the following set of ground rules.

In order to perform fallacy-free analysis of the impact of a proposed change in the allocation of coastal zone resources on the market wealth of a portion of society, any manipulation of numbers must be preceded by four essential steps:

1.  Define the black box:  whose change in income is being analyzed?  Make the definition explicit.
2.  Completely specify all alternatives being considered. Make the specification explicit.
3.  Choose a baseline against which changes in black-box income are to be measured.  Make the baseline explicit.
4.  Set up a complete and consistent set of accounts.

These four pre-analysis steps can be employed as a checklist by officials examining economic analyses put forward by pro-development and anti-development forces.  By merely asking the proponent, "What is your black box?  What is your baseline?" and so on, many of the more obvious errors in the analysis will be exposed.  Better yet, let the proponents know beforehand they will face this checklist.  It may improve the quality of the analyses considerably.

Having performed the four pre-analysis steps, the analysis becomes almost mechanical.

5.  Estimate the yearly income flow into and *out of* each account for each alternative *relative to the baseline*.
6.  For each alternative, aggregate the yearly flows and take present values.
7.  Compare the resulting present values across alternatives. The differences will be the estimated changes in black-box income associated with one alternative rather than another. These differences will then have to be compared with differences in non-market effects.

The principal difficulty associated with Step 5, wherein all the work lies, will be in handling uncertainty.  When estimates depend in an important way on a variable which cannot be predicted with any degree of accuracy, our advice is to work with a range of possibilities.  Assume a low value of the variable and work it through.  Then assume a high value and work it through.  Present the results for a range of cases.  This will in general generate considerably more insight than the much more common practice of using a best guess and tacitly forgetting about the uncertainty.

With respect to Step 7, uncertainty with respect to the interest rate to be used in taking present values will usually make it worthwhile to compute the present values

for a range of interest rates.  In general, uncertainty
should be tackled head-on rather than swept under the rug.
    Some will find the results of such exercises disappoin-
ting.  Even without uncertainty, the dollar figures will not
dictate a particular choice.  After all, real black-box income is but one dimension of a problem.  There will be many
other black boxes, both bigger and smaller than the one
chosen.  In addition, whatever black box we use, the results
ignore non-market effects.  Finally, running analyses over
ranges of uncertain variables will in general result in a
range of results.  Often with one hypothesized value of the
uncertain variable, the number will point to one alternative;
with another hypothesized value of an uncertain variable,
the results will point to a different choice.  One may well
ask, after such an exercise, what have we learned?
    In this regard, the analysis will have to speak for it-
self on a case-by-case basis.  But whatever the application,
the point of the drill is not to dictate a choice.  Rather,
it is to obtain insight into the implications of our choices
with respect to market wealth.  If replacing self-serving,
fallacy-prone estimates of these implications with a range
of fallacy-free estimates affords us these insights, then
we would have to be complete cynics to believe that our decisionmaking with respect to coastal zone resources will not
be improved.